Calculus

Without Hocus Pocus

Notice & Explanatory

Persons attempting to find a motive in this narrative will be prosecuted; persons attempting to find a moral in it will be banished; persons attempting to find a plot in it will be shot. None of the above fates awaits persons who have concluded that only an author allergic to dictionaries would leave the last two words in the title separated by a space.

The first sentence on this page is the text of of the NOTICE in both the first English and the first American edition of the novel whose title is usually (and wisely) shortened to *Huckleberry Finn* and even *Huck Finn*. On the next page in each edition [1, 2], the author has inserted an EXPLANATORY. The present book needs one too, but it will be given on this page.

A dash protruding at the end of the third word on the front cover looked, to my eyes, like a wart at the tip of a well-shaped nose. After considerable pro-and-conning, I decided to act like a cosmetic surgeon and excised the hyphen. In the main text I have toed the line and re-tained the hyphen, except when quoting someone who omitted it (de-liberately or due to a momentary lapse of memory). Searching for other deviants, I came across Kurt Vonnegut's 1990 novel *Hocus Pocus*, but he used the unhyphenated form also in the text. One of my favourites, George Bernard Shaw (ever so finicky about spelling), turned out to be inconsistent. For persons who doubt the veracity of the last statement, the present author has some words of advice: Have faith. Don't verify. Spend the hours so saved on reading *Calculus Without Hocus Pocus*.

With deep respect and affection, I dedicate this book of uncertain genre to two of my colleagues in the Norwegian University of Science and Technology (NTNU):

JØRGEN LØVSETH, who engineered my move from
Zurich to Trondheim;

SIGMUND WALDENSTRØM, who enriched my years in Trondheim by sharing his wondrous skills with me.

Contents

Preamble   ix

What's New Here   xix

Part I: The Nitty-Gritty

1   *The Three Species of Calculus*   3

2   *Integration: Part I*   23

3   *Integration: Part II*   39

4   *Two Variables: Differentiation*   51

5   *Two Variables: Integration*   69

6   *Iterations and Expansions*   81

7   *Differential Equations. Part I*   95

8   *Differential Equations. Part II*   109

9   *Conversion of Operational to Algebraic Forms*   123

Part II: Historical

10 *British Mathematics After Newton* 143

11 *Attempts to Algebrify Calculus* 159

12 *Earnshaw's Pamphlet* 167

13 *Airy's Allies and Adversaries* 207

14 *John West's Posthumous Book* 229

15 *Menger's Version of Calculus* 235

Appendices

A *Differentiability, Differentials and Derivatives* 251

B *Notation and Typesetting* 257

C *The Long and Short of the Integral Sign* 275

D *Quadling's Quandary* 287

E *Jacobi's Jewels* 297

Preamble

The author hopes to ... produce a text-book that can also be read 'for pleasure as well as profit' by those no longer in school; by those who have not been to school; or by those who in their college days suffered those things which most of my own generation suffered.

A private word to teachers and professors will be found toward the end of the volume. I am not idly sowing thorns in their path. I should like to make even their lot and life more exhilarating and to save even them from unnecessary boredom in class-room.

Ezra Pound in *ABC of Reading* [3, p. 11]

Is there *any* hocus-pocus in calculus? A good deal of it, as a matter of fact. Not in calculus per se, but in the teaching of calculus, as anybody who has learnt/taught calculus must know; to disabuse those who have not noticed any problems, some concrete evidence will be presented in support of my claim.

The purpose of *Calculus Without Hocus Pocus* is to shed light on those issues (and only on those issues) which are not treated adequately in standard textbooks of calculus, and to furnish better explanations of those conundrums and paradoxes which have been nagging the minds of students and teachers of calculus for generations. This is not a textbook of calculus in the ordinary, or in any other sense of the term, for it has no set lessons or drills, no problems to be given to a group of learners. Whatever has been omitted may be found in scores of excellent books freely available, and no purpose would have been served by their inclusion.

I have been reading books on calculus and differential equations for close to sixty years. The availability—thanks to the Internet Archive (and other repositories)—of old and rare textbooks during the last dozen or so years has allowed me to consult material that, in former times, would have been beyond the reach of even the most privileged and voracious bookworm, and I daresay, though I may not be able to convince anyone outside the circle of my close friends, that I have read more books (and articles and reviews) on calculus than any other human being, alive or dead. The longer I waited, the more I was able to read and learn about the development of calculus through the ages. Having taken so much from my predecessors, I would like to repay the loan, and it is clear that such repayment should not be deferred indefinitely. Releasing the book sooner would have been premature, by definition; delaying publication much longer would make me a defaulter. I am returning to the world, here and now, whatever I found to be useful and within my grasp, and, like all grateful and conscientious debtors, I feel obliged to return the borrowed capital only after topping it up with a layer of interest. From now on, I will abbreviate the title of the book and call it *Calculus-WHP*.

Calculus-WHP is a series of essayish chapters and appendices, dealing with some crucial aspects of calculus, with the development of calculus itself as the backbone of modern science, and with the conceptual and notational changes that have taken place, or failed to take place, since the middle of the nineteenth century. It is a free discussion of the present-day status of calculus and of the relations of this science to other fields of technical study. The "ins and outs" of calculus, those related ideas that in most modern textbooks are either altogether omitted, or made the subject of fine-print 'boxes' or footnotes, occupy central stage here. The aim has been to produce a condensed and readable book that throws light from various directions upon the difficult parts of this very technical (and somewhat unpopular) subject, and also to show some of the reasons why calculus has been cast in the mould in which we find it. Most of the nontechnical aspects, and even some of the technical issues, have been consigned to Part II and five appendices, which take up almost as many pages as does Part I.

What are the most common conundrums of calculus?

The teaching of differentials has been a topic of lively debate among teachers of calculus; the reader who is interested in knowing the opinions and suggestions of these teachers should consult their contributions in *The Mathematical Gazette* [4–18] and in *The American Mathematical Monthly* [19–25].

At the end of Ransom's 1951 paper in *Amer. Math. Monthly*, the editor (Carl B. Allendoerfer) appended the following note:

> This paper by Professor Ransom is the outgrowth of extended discussion between him and the editor arising from a discussion group at Tufts College in August, 1950 on the "Teaching of Calculus." Although the editor does not agree with Professor Ransom at many points, he is publishing this paper as a stimulus to discussion of an area in which there seems to be great confusion among textbook writers and teachers. Classroom Notes will welcome a limited number of replies to Professor Ransom's paper, particularly those which discuss the questions:
>
> 1. What is a differential?
>
> 2. Does the "dx" of differential calculus have the same meaning as the "dx" in the integral: $\int_a^b f(x)\, dx$?

The editor rounded up the discussion with a disquisitional contribution of his own [26]:

> After reading the numerous papers submitted to Classroom Notes on differentials, and after discussions with other mathematicians, your editor is convinced that there is no commonly accepted definition of differential which fits all uses to which this notation is applied. In the theory of functions of a single variable, $y = f(x)$, a reasonable case can be made for the customary definition: $dx = \Delta x$; $dy = f'(x)\, dx$. This breaks down, however, when one extends it to functions of several variables and considers double integrals of the form $\iint f(x, y)\, dx dy$. Students are rightly baffled when they attempt to convert such an integral to polar coordinates and are told that no longer is it permissible to put $dx = -r\sin\theta\, d\theta + \cos\theta\, dr$, etc. *The Jacobian must be used instead, and at this point the logical structure which was built so carefully collapses entirely.* (My italics.) If we wish to make calculus an intellectually honest subject and not a collection of convenient tricks, it is time we made a fresh start.

Consideration of the uses of differentials in various parts of calculus indicates that mathematicians really do use these symbols to mean quite different things in different circumstances. It seems to the writer that we should admit this from the outset and not try to give a single definition. But surely we must give our students some explanation of differentials or they will be more confused than ever. The following remarks are an attempt to do this. The best description of a differential that I have been able to formulate is: ...

With due respect to Allendoerfer, I must say that the a *carefully built* logical structure would not have collapsed merely by an increase in the number of independent variables from one to two. A logically sound framework cannot be as fragile as the skeleton of the proverbial camel whose back was broken by the addition of the last straw. A fresh start ought to have been made already before the middle of the nineteenth century, when the number of calculus books was comparatively small, and knowledge of calculus was limited to a handful of people, before the time described by Sylvester (1814–97) in a lively language [27, p. 115]: "When I was young, a boy of sixteen or seventeen who knew his infinitesimal calculus would have been almost pointed at in the streets as a prodigy, like Dante, as a man who had seen hell. Now-a-days, our Woolwich cadets at the same age, talk with glee of tangents and asymptotes and points of ...".

Coincidentally with the above debate, Karl Menger published a book, entitled *Calculus: A Modern Approach* [28–30], but the choice *Calculus: A Fresh Start* might have been even more on the mark, for his aim was nothing short of a complete overhaul of the teaching of calculus. Since it is not convenient to discuss the contents of Menger's book at this point, the discussion is postponed.

More than fifteen years after Allendoerfer expounded his understanding of differentials, an exasperated mathematics teacher by the name of Steller addressed his colleagues in a letter which began with the following words [31]: "Many people think that mathematicians when defining new concepts or proving theorems choose their words with care, that ambiguity and inconsistency are anathema to them. Of course, mathematicians strive toward this ideal but, being human, they sometimes fail; yet so strong is the ideal image that when they fail no complaint is raised." A later passage is reproduced below:

Another example of an unfortunate choice of words in English textbooks is "differential coefficient" for derivative. In most (elementary) textbooks it is moreover stressed that, though dy/dx looks deceptively like a quotient, it should never be looked upon as such. This becomes a bit embarrassing when the relation

$$\frac{dy}{dx} = \frac{dy}{dt} \cdot \frac{dt}{dx}$$

must be established. An example of woolly thinking—which is copied continually by many authors of texts on calculus—is the introduction of the integral as the anti-derivative rather than the limit of a sum. The meaning of the term "integral" is lost with this approach and it becomes obviously impossible to show why the fundamental theorem of calculus is so fundamental since it is a direct consequence of the incorrect definition of the integral.

That the problems mentioned above have not stopped confounding beginners would become obvious to anyone who looks at the material available on the Internet, where questions concerning the notation of calculus are regularly raised and discussed. Some of these questions are listed below, the first of which is numbered as 3 because it will be necessary later to refer to the questions raised in the preamble as Q-1. Q-2, &c. .

3. Why is treating the derivative like a fraction an abuse of notation? [32].

4. Can Leibniz notation be treated as a quotient? [33]

5. Is $\frac{dy}{dx}$ not a ratio? [34].

6. How misleading is it to regard $\frac{dy}{dx}$ as a fraction? [35].

7. A really basic integration question concerning differentials [36].

8. Usage of dx in integrals [37].

9. What is dx in integration? [38].

10. Why absolute values of Jacobians in change of variables for multiple integrals but not single integrals? [39]

Here is text that comes after Q-9:

When I was at school and learning integration in maths class at A Level my teacher wrote things like this on the board.

$$\int f(x)\,dx \tag{1}$$

When he came to explain the meaning of the dx, he told us "think of it as a full stop". For whatever reason I did not raise my hand and question him about it. But I have always shaken my head at such a poor explanation for putting a dx at the end of integration equations such as these. To this day I do not know the purpose of the dx. Can someone explain this to me without resorting to grammatical metaphors?

The question elicited several comments and answers. One comment seems worth reproducing: "I sympathize. We need to put a "full stop" to this kind of teaching." The answers left me in no doubt that writing this book would be one step towards putting a half, if not a full, stop to this kind of teaching.

The preface is not the place to pillory those who have given, in their lecture notes or books, puerile answers to the above questions, but the case for yet another tract on calculus might become stronger if I collect some of these answers elsewhere in the book.

Calculus instruction

I became aware of the term "financial meltdown" during one of the many crises that have threatened, over a period very much longer than my lifetime, the economic well-being of the well-to-do, and then subsided after forcing the market to "correct itself", and I am sure that I was not the first one to think of the analogous term "academic meltdown". Ezra Pound, who was not handicapped by a paucity of pungent opinions and caustic words, expressed his thoughts about the teaching of poetry in terms that were not meant to endear him to the Establishment (for which he used a different word). Not all of his words are directly applicable to calculus, and I prefer, because I count a couple of mathematicians among my close friends, not to be as harsh as Pound, but I cannot resist quoting a passage, in which I have replaced his word *deliquescence* by *meltdown* [3, p. 200]:

In general we may say that the meltdown of instruction in any art proceeds in this manner.

1. A master invents a gadget, or procedure to perform a particular function, or a limited set of functions.

   Pupils adopt the gadget. Most of them use it less skilfully than the master. The next genius may improve it, or he may cast it aside for something more suited to his own aims.

2. Then comes the paste-headed pedagogue or theorist and proclaims the gadget a law, or rule.

3. Then a bureaucracy is endowed, and the pin-headed secretariat attacks every new genius and every form of inventiveness for not obeying the law, and for perceiving something the secretariat does not.

It appears to me that the instruction of calculus entered the third phase about a hundred years ago. As far as I can see, the last important change in notation took place in the first decade of the twentieth century, when Leathem introduced a new symbol for the process of taking a limit [40, p. 17]. I quote from the preface to the second edition of his book [41], published in 1922:

> The author's arrow notation for passage to a limit, since its publication in the first edition of this work in 1905, has been adopted by many writers on Pure Mathematics, and may be regarded as now well established. Its application has rightly been confined to continuous passages to limit, and there is evidently room for some corresponding symbol to indicate saltatory approach to a limit value. *saltation*: leap, A dotted arrow might perhaps appropriately serve this purpose; sudden it would present no difficulty to the printer, but it is just doubtful transition whether it would be convenient in manuscript work.

Among those who supported Leathem's proposal was G. H. Hardy, who showed his approval in the preface to the first edition of his highly acclaimed book *A Course of Pure Mathematics* [42].

The examples given in this book will show the reader that, during the nineteenth century, the notation used by mainstream mathematicians changed substantially, but the Leibnizian symbol $\int \ldots dx$ had already become a sacred cow by the end of the

previous century, and the faceless "bureaucracy" and " secretariat" handling the affairs of calculus have displayed, more recently, an aversion to notational and nomenclatory innovation
compared to which the orthodoxy of the most fundamentalist
priests of the Abrahamic religions is a tiny fraction of an epsilon,
a mere infinitesimal, one might say. The deep conservatism of
the calculus priesthood is understandable and, on the whole, justified, and the results, or most of them, speak for themselves.
However, a public library that does not throw out, on a regular basis, some of its books to make room for new arrivals, must
eventually lose the interest of the public and their patronage.
Calculus, the mathematics of change, cannot be exempt from
change. A mathematical discipline is bound to atrophy if its
practitioners stubbornly refuse to make any changes in their notational scheme.

Writing a book seems to be the only opportunity for proposing some changes which, in my opinion, are sorely needed, and
none of which is too big to cause convulsions.

Calculus reform

As might be expected, there has been much talk of calculus reform during the last few decades [43–46]. One or two brave souls
have gone beyond mere talking, and tried to take the bull by
the horns, but these noble matadors were so brutally gorged by
those savage horns that other would-be reformers have been dis-
couraged from entering the bullring. I am tempted to misquote
Pound once more, but on this occasion, I have made only one
change, and substituted *calculus* for *moral* [3, p. 158]:

> There is nothing to be said against calculus reform. Born in
> a town with bad sewers, the man with a good nose will certainly
> agitate for their improvement. It is not the pleasantest occupation,
> nor the highest use of human faculties.
> But the man who agitates is an infinitely better fellow than the
> parasite who sabotages the work, or who waits till he can get a
> percentage on the contract for new cloacae.

cloaca (pl. *-ae*):
sewer

Why has all this talk of reform failed to change the teaching of
calculus? Things would have been different if we really did have

a secretariat of calculus, and if this secretariat had agreed to undertake a major overhaul. Murray H. Protter, a respected mathematician and coauthor of a successful calculus book, wrote an opinion piece entitled "Calculus Reform" in *Mathematical Intelligencer* [43]. The introductory part of his essay is followed by six sections bearing the titles: 1) How a Student Studies Calculus, 2) Growth in the Size of Textbooks, 3) Publishers, 4) Applications, 5) Calculus and Computers, 6) Conclusions. Those who have been involved with teaching mathematics, if asked to write an essay structured around the first five headings, would produce something whose contents are not likely to differ substantially from Protter's essay. Their conclusions may differ, however. I reproduce the last paragraph of Protter's essay:

> Finally, how do we change the study habits of the average student? Or should we try? The organization of most texts works against change; unless we develop another system we will be locked into the one we have.

The system *we have* has been in existence for over two centuries. It is based on instruction (mostly lecturing) and examination. We have not been able so far to devise a system for assessing the mathematical skills of a large number of students in a short time that is more efficient and fairer than an examination in which the candidate is asked to answer questions set by the examiner(s). If a student is perceptive enough to discern that the examination papers follow a pattern, and that it is possible to predict most of the questions that (s)he will have to answer, one should blame, not the student for having prepared well, but the examiner for asking questions that could be easily anticipated. Under the current system, where no kudos is attached to setting examination papers that do not rehash questions from earlier papers, there is no reason to expect a change in the attitude of a candidate.

Resistance to reforms

The trouble with reforms is that even those who perceive the benefits of the proposed changes may oppose it for partisan reasons. Take, as an example, the calendar reform introduced by Pope Gregory XIII. Everyone who had the knowledge to grasp the technical aspects of the reform knew that it would be foolish to oppose it, but opposition there was. How long do you suppose it took the opponents to come to their senses? Nearly two hundred years!

Alexander Philip writes [47, p. 22]: "The year 1582 was the initial year of the Gregorian Calendar, which was at once adopted by the various countries which recognised the spiritual authority of Rome. France adopted the new style in December, 1582. Switzerland, the Catholic Netherlands and the Catholic States of the Empire in 1583. The Protestant States for a considerable time refused to follow. In 1699, however, chiefly at the instigation of the philosopher Leibniz, the Protestant States of Germany came into line. In Great Britain the new style was not adopted until the passing of the Calendar New Style Act (1750), under which Act it came into operation in 1752."

The 'philosopher Leibniz' was also a mathematician and professional diplomat. He was instrumental in establishing scientific academies in Berlin, St. Petersburg, Dresden, and Vienna. In 1682 he founded the journal *Acta Eruditorum*, in which he subsequently printed his discovery of the differential calculus.

Even the Gregorian calendar does not satisfy those who are in favour of a more rational version called the Perpetual Calendar or the World Calendar, but more than a century of persuasion has not succeeded in dislodging the Gregorian calendar [48–50].

At this point I am tempted to revert to Pound's analogy of bad sewers. There is little hope, and really no need, for installing a new sewage system in the realm of calculus. In a long career spent largely in making (and sometimes correcting) my own blunders, I could hardly avoid noting and correcting the mistakes of previous authors, but I could detect and deodorize only the malodorous lumps which offended my peculiar intellectual olfactory sense. The remainder awaits the attention of those blessed with better noses and brains.

What's New Here

I can hardly expect that a mathematician, to whom this subject is familiar, will look through the whole work to pick out here and there a theorem, or a mode of proceeding, which has some point of novelty. I therefore subjoin references to those parts of the work for which I have not been indebted to my knowledge of what has been written before me: much of what is cited is probably not new, indeed it is dangerous for any one at the present day to claim anything as belonging to himself; several things which I once thought to have entered in this list have been since found (either by myself, or by a friend to whom I referred it) in preceding writers.

Augustus De Morgan in the Preface to *The Differential and Integral Calculus* [51, p. v]

SINCE the remarks quoted above were true already in 1842, the word 'new' in the title of the chapter must be interpreted, not as 'new to the world' but as 'likely to be new to the reader', unknown not just to the average or typical reader, but even to those who have been teaching calculus for years, the atypical reader who has been worrying about the issues discussed here—the calculus cognoscenti, so to speak. This additional preface (or advertisement) is written to enable such people to decide whether they should lay the book aside or keep reading a little while longer.

The three species of calculus

The singular and plural forms of the noun *species* are the same, and in everyday usage, the word means *kind* or *sort*.

Until fairly recently, most people used to think that there were only two species of elephants, the African and the Asian. A few sceptics questioned this simplistic view and favoured a greater diversity in the African elephant population, but convincing evidence that there indeed are two species of African elephants only came along in 2001 [52], bringing a century-old debate to a conclusion.

An introductory article on "The Three Species of Elephants" must explain, through words and images and other necessary aids, the characteristic features of the three species in sufficient detail to enable its readers to distinguish a member of one species from that of another, so that its readers may acquire updated information and pass it on to anyone else who is curious to know something about elephants.

As with elephants, so with calculus. Most of those who learn calculus tend to think that there are just two versions of calculus, the calculus of differentials and the calculus of derivatives; for convenience, these versions have been named in this book as *d*-Calculus and *D*-Calculus, respectively. Unlike the different species of elephants, which do not interbreed, the two species of calculus produced a hybrid, named H-Calculus in this book, which has come to dominate the teaching of calculus.

Chapter 1, whose scope is similar to that of the above named fictitious article on elephants, explains the similarities of the three species of calculus as well as their dissimilarities; most of the questions mentioned in the Preamble would not have arisen if the differences had not been blurred.

De-extinction of the woolly mammoth

The idea of reviving species that have become extinct is called de-extinction. There has been much talk, accompanied by some action, of bringing back to life some of the extinct animals, including the woolly mammoth. Is this a good idea? Some extinct animals did play vital roles in their ecosystems, and ours might benefit from their de-extinction [53]. For much the same reason, I have resurrected two symbols which have become extinct. Since it is not convenient to mention both at this point, I content myself by recalling the symbol

$$\int_x f(x) \qquad \text{(S-1)}$$

proposed by Airy (and used by a small group of his Cambridge colleagues) for an antiderivative of $f(x)$. The symbol $\int_x$ is equivalent to, but uses fewer characters than, the more familiar notation D_x^{-1}. The corresponding symbol in H-Calculus is

$$\int f(x)\, dx, \qquad \text{(S-2)}$$

and the presence of dx is justified by saying that it indicates the variable of integration. Although this is somewhat like sticking the large ears of an African savanna elephant on the ears of an Asian elephant and hoping that no one will get confused, one need not quibble over the oddity of the notation as long as dx is not treated as a differential. Although this caution is frequently ignored when integration by substitution is carried out with the help of the seemingly obvious formula

$$\int f(x)\, dx = \int f[\phi(u)] \frac{dx}{du}\, du, \quad x = \phi(u), \qquad \text{(2)}$$

a meticulous procedure that avoids this solecism (at the cost of a few more mathematical steps) can be used to assure students that all is well.

Change of variables in a double integral

In the traditional approach to changing variables in a double integral, say from a pair of old variables (x, y) to two new variables (u, v), $dxdy$ is treated as a differential area element, equal to the product of two differentials dx and dy, and the student is told that the (u, v)-counterpart of the cartesian area element is not $dudv$, but $|J|dudv$, where J is the relevant Jacobian determinant. This makes complete sense in d-Calculus, but not in H-Calculus. A student of H-Calculus who has grasped that Eq. (2) is a consequence of the chain rule (to a single change of variable), if encouraged to extend the reasoning to the two-dimensional case, is likely to surmise that the recipe for the double change would take the following form:

$$\iint f(x,y)\,dxdy = \iint f[\phi(u),\psi(v)] \left(\frac{\partial x}{\partial u}\right) \left(\frac{\partial y}{\partial v}\right) dudv. \quad (3)$$

Let us note in passing that, if one uses Airy's notation, the above formula would be written as

$$\int_x\int_y f(x,y) = \int_u\int_v f[\phi(u),\psi(v)] \left(\frac{\partial x}{\partial u}\right) \left(\frac{\partial y}{\partial v}\right). \quad (4)$$

Even those who are not able to fabricate on their own the half-baked formula shown in Eq. (3), would not be as perplexed by this formula as by the recipe involving the Jacobian.

Four full-fledged versions of Eqs. (3) and Eq. (4), showing which variable is to be kept constant in each partial derivative, are derived in this book with the aid of *a more elaborate notation for partial differentials*; the absolute value of each product of partial derivatives in the transformed integral can be shown, if one wishes to do so, to be equal to $|J|$. Once the notation is adopted, other formulas involving partial derivatives, including those which are needed in thermodynamics, can be obtained by using elementary algebra; no prior knowledge of Jacobian determinants is needed in these manipulations.

The need for a nuanced notation in many-variable calculus was noted by Jacobi. For a function $f(x,y)$ of two independent variables x and y, he suggested not only the symbols $\dfrac{\partial f}{\partial x}$ and $\dfrac{\partial f}{\partial y}$, but also

$$\int f(x,y)\,\partial y, \quad (5)$$

because x is to be kept constant when the integration over y is performed. On going back to Eq. (3), one sees that the equation would take a more symmetric appearance if d was replaced throughout by ∂.

A symbol with more than one meaning

It will be convenient to use, when it seems desirable to do so, the abbreviations anti^D and anti^d for the terms *antiderivative* and *antidifferential*, respectively.

The symbol

$$\int f(x)\, dx$$

is usually interpreted, in H-Calculus, as the most general anti^D of $f(x)$; in d-Calculus, as the most general anti^d of $f(x)\, dx$. In this capacity, it stands for an infinite set of functions. However, this interpretation is not convenient when one discusses the general solution of a differential equation; in this context, the symbol is used to represent a *particular* anti^D or anti^d.

In the past a super-scripted integral sign, like one of those shown below

$$\int^x f(x)\, dx = \int^x f(u)\, du = \int^x f(v)\, dv$$

has sometimes been used to denote an anti^D (or anti^d) to which no constant (arbitrary or otherwise) has been *purposely* added. Since other interpretations have also been placed on this symbol, it is desirable to use a different superscript, and the alternative shown below

$$\int^\dagger f(x)\, dx \tag{6}$$

has been used here.

In the parlance of differential equations, the integral in (6) is a particular integral of the equation

$$\frac{dy}{dx} = f(x), \tag{7}$$

and the arbitrary constant is the complementary function (the general solution of $dy/dx = 0$, the homogeneous equation obtained by setting the right-hand side of Eq. (7) to zero).

Linear ODEs with constant coefficients

The operators D and D^{-1} have become standard items in the toolkit of those who take a first course in calculus. I suggest that the operators $(D + a)$ and $(D + a)^{-1}$ (where a is a constant) should be added to this kit, and they should be made aware that, for several familiar and important functions, $(D + a)^{-1} f$ can be inferred from the expression for $(D + a)f$. They should also be informed that the operator $(D+a)^{-1}$ facilitates the calculation of certain integrals; indeed the symbolic method provides the easiest way of handling the awkward integral

$$\int e^{-x} x \sin x \, dx.$$

The operator $(D+a)$ arises naturally when one considers the first-order ODE

$$\frac{dy}{dx} + ay = f(x). \tag{8}$$

This book stresses the fact (known for two centuries but carefully hidden from learners) that anyone who is able to solve the above equation can also solve, without acquiring new skills, the second-order ODE

$$\frac{d^2y}{dx^2} + 2b\frac{dy}{dx} + c^2y = f(x), \tag{9}$$

refrain (n): which, everyone will agree, is the refrain common to every course
recurring in science and engineering. The use of the operator $(D+k)^{-1}$
phrase or makes light work of solving such equation by providing a single
line, recipe for solving constant coefficient linear ODE's.
especially at
the end of
stanzas

Necessity obliges us to neologize

The noun *neologism* means a newly coined word (or phrase or doctrine) or the use of old words in a new sense. I have tried to follow Thomas Jefferson's advice, given in a letter to John Waldo [54, pp. 184–5]: "I am no friend, therefore, to what is called *Purism*, but a zealous one to the *Neology* which has introduced these two words without the authority of any dictionary. I consider the one as destroying the nerve and beauty of language, while the other improves both, and adds to its copiousness." The phrase which forms the title of this section is taken from the same letter [54, p. 189].

I felt the need to introduce the modifiers *active* and *passive* for distinguishing the variable with respect to which a function of two or more variables is differentiated or integrated from the variable (or variables) which is (are) held fixed during the operation being performed. If my neologisms are not hopelessly inapt, the reader will be able to infer which modifier refers to which variable. This rather coy attitude will be abandoned when the two terms are introduced and used In Chapters 4 and 5.

The term *integral* is a synonym for *antidifferential* in *d*-Calculus and for *antiderivative* in *D*-Calculus. Another, less known synonym is the term *primitive* (used as a noun). There is a need for a verb that could serve as the counterpart of *integrate*, and I propose to use *primitate* for 'find a primitive of'; *primitation* can serve as the counterpart of 'antidifferentiation'.

Usually $D_x y = dy/dx$ is called the derivative (or differential coefficient) of y with respect to x. It will prove convenient to use instead the phrase "x-derivative of y"; likewise $D_x^{-1} y = \int_x y$ will be called an "x-antiderivative of y" or an "x-primitive of y".

Part I: The Nitty-Gritty

Chapter 1

The Three Species of Calculus

Of course I shall assume that you all know how to differentiate and integrate, and that you have frequently used both processes. We shall be concerned here solely with more general questions, such as the logical and psychological foundations, instruction, and the like [p. 207].

Consequently he [Lagrange] talks of *derivative calculus* instead of *differential calculus* [p. 220]. (Emphasis added.)

Felix Klein [55] .

FIRST there was *differential calculus*, then came *derivative calculus*, and after that came a deluge of confusion, a century or more of notational promiscuity, nomenclatorial anarchy, pedagogical pettifogging, and a cornucopia of calculus books (amongst them many jewels), most of which are available at no cost to anyone who wants to learn calculus.

The calculus which has been taught during the last 10–15 decades is a hybrid of two distinct species, which will be called here *d*-Calculus and *D*-Calculus, and the name H-Calculus will be used for the hybrid itself. I assume that my readers, like those of Klein, have been using H-Calculus, which means that the time has now come to examine the similarities and differences between the progeny and its two progenitors.

d-Calculus: differentials, integrals and differential coefficients

To begin with a comment concerning the notation to be followed henceforth: When quoting from a source where an upright font

3

is used for writing the first letter of a differential (for example dx or dy), I will use a slanted letter and write dx, dy, etc.

Since the names and symbols introduced by Leibniz have become mathematical icons, it will be sufficient to say that the label d-Calculus will be applied in this book to the Leibnizian version, where differentiation begets differentials, and these entities are denoted by symbols like dx and $dF = fdx$; for what we would call an *antidifferential* (or anti^d for short), Leibniz and his followers used the name *integral* and the symbol $\int$. The operations of differentiation and antidifferentiation (or integration) are interrelated as follows:

$$d \int dF = dF, \quad \int f\, dx = F(x) + C, \tag{1.1}$$

where C is a constant in the sense that $dC = 0$.

When convenient, I will use d^{-1} as an alternative to $\int$. Integration is an operation, and the corresponding operand is called the integrand. In d-Calculus, the integrand is always a differential (see Box 1.1).

Box 1.1. Symbol for antidifferential in d-Calculus

$$\underset{\text{integral sign}}{\longrightarrow} \int \overset{\text{integrand}}{\overbrace{f(x)\, dx}}$$

In a relation of the type $dF(x) = f(x)\, dx$, which connects two differentials, their quotient

$$\frac{dF(x)}{dx} \tag{1.2}$$

is called the *differential coefficient* of $F(x)$ with respect to x.

D-Calculus: derivatives and antiderivatives

If $y = F(x)$ and $y + \Delta y = F(x + \Delta x)$, the *derivative* of y with respect to x (or simply the x-derivative of y) is denoted by the symbol $D_x y$ or $F'(x)$, and defined as follows

$$D_x y \equiv \lim_{\Delta x \to 0} \frac{\Delta y}{\Delta x} = \lim_{\Delta x \to 0} \frac{F(x + \Delta x) - F(x)}{\Delta x}. \tag{1.3}$$

With $f(x)$ denoting the x-derivative of $F(x)$, any function whose x-derivative equals $f(x)$will be called an x-antiderivative of $f(x)$, and any such antiderivative can be expressed as $F(x) + C$. I will often use the abbreviation anti^D for the term *anitderivative*.

It has become customary to use the symbol $D_x^{-1} f(x)$ for an anti^D of $f(x)$, and this custom will be followed here; however, I will also use the alternative shown in Box 1.2; the sign $\int_x$ was proposed by Airy [56], who called it 'symbol for integration'. A third option, due to West [57], is explained in Chapter 14.

Box 1.2. Airy's symbol for integration (in D-Calculus)

integrand

integral sign $\longrightarrow \int_x f(x)$ $\longleftarrow$ variable of integration

Using the symbol $\int_x$, we may write the D-Calculus analog of Eq. (1.1) as

$$D_x \int_x f(x) = f(x), \qquad \int_x D_x F(x) = \int_x f(x) = F(x) + C. \quad (1.4)$$

The mathematicians who have adhered strictly to D-Calculus can be counted on the fingers of one hand, and the number of books where integration is handled without mentioning differentials is even smaller. The topic of integration without differentials is discussed at greater length in Chapters 2, 13, and 15.

H-Calculus: a potpourri of symbols

potpourri: a mixture or medley

The first three decades of the twentieth century were the best of times and the worst of times for physics. Light, which was understood to be a wave phenomenon, seemed to masquerade occasionally as a stream of particles; conversely, electrons, the first subatomic corpuscle (particle), sometimes displayed wave-like properties. The physicist William Bragg described the confusion in the following words [58]:

> For the present we have to work on both theories. On Mondays, Wednesdays, and Fridays we use the wave theory; on Tuesdays,

Thursdays, and Saturdays we think in streams of flying energy quanta or corpuscles.

The situation in H-Calculus is nearly as murky. Symbols that have no meaning (or a particular meaning) before the introduction of differentials acquire, respectively, a meaning (or some other meaning) afterwards. The answers to Q-3–Q-9 in the Preamble cannot be given until we know which week of the semester we are in and which day of the week it is.

In H-Calculus, the derivative has the same definition as in D-Calculus, but the architects of H-calculus chose to retain not only the Leibnizian *symbol dy/dx* but also the *name* differential coefficient for the derivative.

G. H. Hardy is an author to whom I pay utmost respect and all the attention I can muster. Chapter VI of *A Course in Pure Mathematics* is entitled 'DIFFERENTIATION AND INTEGRATION, and here Hardy writes [59, p. 205]:

> We have already explained that what we call a *derivative* is often called a *differential coefficient*. Not only a different name but a different notation is often used; the derivative of the function $y = \phi(x)$ is often denoted by one or other of the expressions
>
> $$D_x y, \quad \frac{dy}{dx}.$$
>
> Of these the last is the most usual and convenient: the reader must however be careful to remember that dy/dx does not mean 'a certain number dy divided by another number dx': it means 'the result of a certain operation D_x or d/dx applied to $y = \phi(x)$', the operation being that of forming the quotient $\{\phi(x+h) - \phi(x)\}/h$ and making $h \to 0$.

The writer of the above paragraph knows that an explanatory remark is sorely needed; as one disgruntled teacher has pointed out [60], some thoughtful students are bound to ask 'if dy/dx is not a quotient, as you keep telling us, why do you write it as though it is?' Hardy has an answer [59, pp. 205–6]:

> Of course a notation at first sight so peculiar would not have been adopted without some reason, and the reason was as follows. The denominator h of the fraction $\{\phi(x+h) - \phi(x)\}/h$ is the difference of the values $x + h$, x of the independent variable x; similarly the numerator is the difference of the corresponding values

$\phi(x+h), \phi(x)$ of the dependent variable y. These differences may be called the *increments* of x and y respectively, and denoted by δx and δy. Then the fraction is $\delta y / \delta x$, and it is for many purposes convenient to denote the limit of the fraction, which is the same thing as $\phi'(x)$, by dy/dx. But this notation must for the present be regarded as purely symbolical. The dy and dx which occur in it cannot be separated, and standing by themselves they would mean nothing: in particular dy and dx do not mean $\lim \delta y$ and $\lim \delta x$, these limits being simply equal to zero. The reader will have to become familiar with this notation, but so long as it puzzles him he will be wise to avoid it by writing the differential coefficient in the form $D_x y$, or using the notation $\phi(x), \phi'(x)$, as we have done in the preceding sections of this chapter.

In Ch.VII, however, we shall show how it is possible to define the symbols dx and dy in such a way that they have an independent meaning and that the derivative dy/dx is actually their quotient.

We are now able to provide a two-part answer Q-3–Q-6 in the Preamble:

1. Prior to the introduction of differentials, the symbol dy/dx is simply a five-character cluster that is meant to be indecomposable. For the definition of the symbol, see Hardy's passages quoted above, or refer to Eq. (1.3) on p. 4, where the symbol $D_x y$ has been used because it does not evoke the idea of a fraction, and its meaning does not depend on whether or not the author intends to introduce differentials.

2. After the introduction of differentials, the symbol dy/dx represents a fraction in which the divisor and the dividend are both differentials, linked by the relation $dy = D_x y\, dx$.

We see from the last paragraph of the second excerpt from Hardy that he discusses integration before introducing differentials. To examine the consequences of Hardy's choice, let us first see how he describes integration [59, p. 230]:

If $\psi(x)$ is the derivative of $\phi(x)$, then we call $\phi(x)$ an *integral* or *integral function* of $\psi(x)$. The operation of forming $\psi(x)$ from $\phi(x)$ [sic: read "forming $\phi(x)$ from $\psi(x)$"] we call *integration*.

We shall use the notation

$$\phi(x) = \int \psi(x)\, dx.$$

8

It is hardly necessary to point out that $\int \ldots dx$ like d/dx must, at present at any rate, be regarded purely as a symbol of operation: the $\int$ and the dx no more mean anything when taken by themselves than do the d and dx of the other operative symbol d/dx.

The phrase "at present at any rate" led me to expect (in or after Ch.VII) some clarifying comments about the presence of the dx in $\int(\cdots)\,dx$, but I could not find any. Still, we can come to one important conclusion in the light of whatever we do find in Hardy's book, where differentials appear on p. 280, and the term *indefinite integral* is defined on p. 287: "On the other hand the integral function

$$F(x) = \int f(x)\,dx$$

is sometimes called the **indefinite integral** of $f(x)$." In the next but one paragraph, we are told: "But when we are considering 'indefinite integrals' or 'integral functions' we are usually thinking of *a relation between two functions*, in virtue of which one is the derivative of the other". On this basis we may conclude that Hardy uses the symbol $\int f(x)\,dx$ to designate an anti^D of $f(x)$ not only before the introduction of differentials, when dx has no independent existence and no meaning, but also afterwards, when dx must be interpreted as a differential.

We will now look at the interpretation given in a book whose authors introduce the symbol $\int f(x)\,dx$ after defining differentials [61, p. 270]: "We think of the combination $\int \ldots dx$ as a single symbol: we fill in the "blank" with the formula of the function whose antiderivative we seek. We may regard the differential dx as specifying the independent variable x both in the function $f(x)$ and its antiderivatives."

Box 1.3. Symbol for antiderivative in H-calculus

integrand

integral sign $\longrightarrow$ $\int f(x)\,dx$ $\longleftarrow x$ is the variable of integration

The string of symbols $\int f(x)\,dx$, which stands for an anti^D of $f(x)$ in H-Calculus (both before and after dx acquires the status of a differential), is visually identical with the symbol for an anti^d of $f(x)\,dx$. No one can tell, merely by looking at the string, whether it is to be parsed according to the scheme shown in Box 1.3 or that in Box 1.1. Of course, the choice of the parsing scheme cannot affect the result, for a function that is an anti^D of $f(x)$ is also an anti^d of $f(x)\,dx$.

parse: resolve into component parts and describe them

Questions Q-7–Q-9 in the Preamble (p. xiii), raised by students who were not satisfied with the explanation given to them about the presence of dx in the symbol $\int f(x)\,dx$, can be answered now: The symbol dx is merely a variable-indicator if the author is yet to introduce differentials; otherwise, it is a differential as well.

It stands to reason that, in the parlance of H-calculus, the term *integrand* should be applied to $f(x)$, because one is now seeking an anti^D of the function $f(x)$. Using the symbols introduced above, we can write

$$\int f(x)\,dx = \int_x f(x) \tag{1.5}$$

$$= D_x^{-1}f(x) \tag{1.6}$$

and we should note the absence of dx on the right hand sides of Eqs. (1.5) and (1.6). Differentiation of (1.5) gives

$$D_x \int f(x)\,dx = f(x) = D_x D_x^{-1}f(x). \tag{1.7}$$

In this case, D_x annihilates not only the integral sign but also the dx, which is to be expected, since neither $\int$ nor dx is supposed to have a separate existence, and differentiation of an indefinite integral must leave us with only the integrand. This can be rendered more transparent by placing dx immediately after the integral sign and restating the result of differentiation as

$$D_x \int dx\, f(x) = f(x) \quad \text{or} \quad D_x \int dx = 1 = D_x D_x^{-1}. \tag{1.8}$$

A matter of mathematical syntax

Putting the integral sign ($\int$) and the differential element (dx) at the two ends of an expression is often compared to the use of parentheses, and it also explains why a teacher suggested that the dx plays the same role as the full stop at the end of a sentence.

The integrand in d-Calculus (see Box 1.1), being a product of two factors, may also be written as $dx\,f(x)$. Within the framework of d-Calculus, it often proves convenient to write

$$\int dF = \int dx\,f(x),\qquad(1.9)$$

and we will see that Fourier, who used d-Calculus, frequently placed dx immediately after $\int$. It is clear that $\int$ and dx side by side even in H-Calculus. One can conclude, by taking Gregory as an exponent of H-Calculus and looking at the list of formulas in his collection, that he wrote $\int dx\,f(x)$ more often than not. To show just one example, I reproduce his formula for *integration by parts* [62, p. 253]:

$$\int dx\,u\frac{dv}{dx} = uv - \int dx\,v\frac{du}{dx}.$$

De Morgan did not hesitate to suggest the following compact notation for an n-fold integral (see p. 266):

$$\left(\int dx\right)^n f(x).\qquad(1.10)$$

How to stop causing confusion

The presence of dx in the anti^D symbol (Box 1.3) causes untold confusion without conferring any authentic advantage; whatever benefits arise are those that can be justified only when dx is a genuine, stand-alone differential. There are at least two effective ways to dispel the confusion, the first of which is likely to ruffle fewer teachers of calculus.

The simplest option

The simplest remedy is to introduce differentials before embarking upon integration, and to insist on integrating only differentials, which amounts to adopting not only the notation but also the spirit of d-Calculus (Box 1.1). Since an anti^d of $f(x)\,dx$ is also an anti^D of $f(x)$, the result cannot change.

As a representative of the authors who have advocated the above standpoint, I choose H. B. Phillips [63, p. 74]:

It should be noted that we always integrate a differential and not a derivative. There are several reasons for this. A very important one is the fact that the differential of a quantity is independent of the variable in terms of which it is expressed. If the same quantity y is expressible in terms of two different variables

$$y = f_1(x), \quad y = f_2(t), \tag{1.11}$$

its differential has equal values whether the one or the other of these expressions is used. But the derivatives of y with respect to x and t are not in general equal. If integration were defined as the process of finding a function with given derivative, there would be a different integral for every variable.

This remark explains why a suffix is needed to specify the variable of integration in the symbol D_x^{-1}, but the symbol d^{-1} needs no such suffix; however, we shall see later (pp. 30–31) that this divide is not unbridgeable.

A different option—inadvisable if used on its own

An alternative strategy is to drop the confounding symbol dx and use one of the symbols on the right hand side of Eqs. (1.5) and (1.6). Some fundamental antiderivatives in Airy's notation are listed in Box 1.4; clearly, an arbitrary constant can be added to each anti^D shown on the right hand side of any formula.

Box 1.4. Some fundamental antiderivatives

$$\int_x \frac{1}{x} = \ln|x| \qquad\qquad \int_x x^n = \frac{1}{n+1}x^n \text{ if } n \neq 1$$

$$\int_x e^x = e^x \qquad\qquad \int_x a^x = a^x / \ln a$$

$$\int_x \sin x = -\cos x \qquad\qquad \int_x \cos x = \sin x$$

$$\int_x \tan x = \ln|\sec x| \qquad\qquad \int_x \cot x = \ln|\sin x|$$

$$\int_x \sec^2 x = \tan x \qquad\qquad \int_x \csc^2 x = -\cot x$$

The ideal solution

Anyone who shares Hardy's opinion about the rewards of using the differential notation will not be satisfied with the second option. However, one is under no compulsion to choose only one of the two alternatives presented above. I know of only one author, viz. W. E. Byerly, who uses both D-Calculus and d-Calculus, without ever hybridizing the two.

Byerly's *Elements of the Differential Calculus* was "intended for a text-book, and not for an exhaustive treatise" [64, p. iii]. Using the symbols D_x and $\int_x$, Byerly covers, in fewer than 300 pages, not only what the title promises, but also a good deal of integral calculus. An updated version of this volume could easily outrival many modern textbooks. In the Preface, Byerly listed some notable features of his book:

> Its peculiarities are the rigorous use of the Doctrine of Limits as a foundation of the subject, and as preliminary to the adoption of the more direct and practically convenient infinitesimal notation and nomenclature ; *the early introduction of a few simple formulas and methods for integrating*; a rather elaborate treatment of the use of infinitesimals in pure geometry; and the attempt to excite and keep up the interest of the student by bringing in throughout the whole book, and not merely at the end, numerous applications to practical problems in geometry and mechanics. (My italics.)

Immediately after introducing differentials, Byerly goes on to inform the reader that the derivative will thenceforward play second fiddle [64, Chapter 11]:

> 180. The differential notation has the advantage over the derivative notation, that it is apparently simpler, and that the formulas in which it is used are more symmetrical than those in which the other notation is employed; and although the differential is defined by the aid of the derivative, and the formulas for the differentials of functions are obtained from the formulas for the derivatives of the same functions, *there is a practical advantage*, after the formulas have once been obtained, *in regarding the differential as the main thing, and looking at the derivative as the quotient of two differentials.* (Emphasis in the original.)

A companion volume entitled *Elements of the Integral Calculus* appeared in 1881 [65], followed by a second edition in 1888 [66]. After the first two chapters, which are of a preliminary nature, the third chapter, titled "General Methods of Integrating", begins with Article 49. In quoting the text of this article below, the notation will be modernized; this means that all dots between various symbols will be dropped and $f.x$ will be replaced by $f(x)$; the reader should note that Byerly refers to the earlier book on differential calculus as I.

> 49. We have defined the *integral* of any function of a single variable as *the function which has the given function for its derivative* (I. Art. 53); we have defined a *definite integral* as the limit of the sum of a set of differentials; and we have shown that a definite integral is the *difference between two values of an ordinary integral* (I. Art. 183).
>
> Now that we have adopted the differential notation in place of the derivative notation, it is better to regard an integral as the inverse of a *differential* instead of as the inverse of a *derivative*. Hence the integral of $f(x)\,dx$ will be the function whose differential is $f(x)\,dx$; and we shall indicate it by $\int f(x)\,dx$. In our old notation we should have indicated precisely the same function by $\int_x f(x)$; for if the derivative of a function is $f(x)$ we know that its differential is $f(x)\,dx$. (Emphasis in the original.)

Essentially the same stance was taken by E. B. Wilson, the last research student of J. W. Gibbs [67]. Wilson devoted the first chapter of his *Advanced Calculus* [68] to a recapitulation of the fundamental rules, and presented within a span of 24 pages, an admirably succinct account of differentiation and integration, in which he denoted an anti^D of $f(x)$ by the symbol $\int f(x)$ and added a subscript only when it was important to specify the integration variable. He added a footnote [68, p. 17], the text of which included the following remark: "The use of this notation is left optional with the reader; it has some advantages and some disadvantages". He did not spell out the merits and demerits of the notation, and chose to employ, without offering any explanation, the notation of d-Calculus in the rest of the book. Most notably (and regrettably, I might add), Wilson assented to the practice of using the symbol dy/dx as a substitute for D_{xy} (the derivative of y with respect to x) before the introduction of differentials.

14

Defining differentials

In a brief note on the teaching of differentials, written in 1950, Hadamard wrote [69]: 'In my opinion, the notion of differential, as formulated by the founders of Infinitesimal Calculus, must be considered as obsolete. As Poincare has already said, "We must think in derivatives."' Hadamard's claim about the obsoleteness of the *original* concept of differentials remains valid. Although the concept was subsequently placed (by Abraham Robinson) on a new foundation, called non-standard analysis, I am inhibited, by my lack of intimacy with mathematical logic, from passing a judgement, or summarizing the new approach; the interested reader is referred to two works by Keisler [70, 71], or a short note by Stewart [72, pp. 518–23]. Within the framework of H-Calculus, the discussion of differentiation starts with derivatives, and differentials can only be defined afterwards.

Let the function $y = f(x)$ have a derivative $D_x f$ at the point x. The increment Δx in the independent variable may be large or small, to be selected at one's pleasure. The differential of the dependent variable, dy, is then defined as:

$$dy = D_x f \cdot \Delta x = f'(x) \cdot \Delta x. \tag{1.12}$$

In particular, if $y = x$, so that $D_x y = 1$, we find that

$$dx = \Delta x. \tag{1.13}$$

This means that dy, the differential of the dependent variable y can be expressed as

$$dy = D_x f \cdot dx. \tag{1.14}$$

Having defined differentials that obey the relation

$$dy \div dx = D_x y, \tag{1.15}$$

we may now equate $D_x y$ with an actual quotient of two finite quantities. The foregoing is true if x is the independent variable. If x and y are both functions of another variable t,

$$x = \chi(t), \quad dx = D_t \chi \, dt, \tag{1.16a}$$
$$y = \eta(t), \quad dy = D_t \eta \, dt, \tag{1.16b}$$

Dividing the second equation in (1.16b) with the corresponding equation in (1.16a), one gets

$$dy \div dx = \frac{D_t \eta}{D_t \chi} = \frac{D_t y}{D_t x}. \tag{1.17}$$

But, in virtue of the chain rule (see below), $D_t y$ can be written as:

$$D_t y = D_x y \cdot D_t x \tag{1.18}$$

We return to (1.17) and continue:

$$dy \div dx = \frac{D_t y}{D_t x} = \frac{D_x y \cdot D_t x}{D_t x} = D_x y. \tag{1.19}$$

The last result allows us to state the following important theorem: *The quotient dy/dx is the derivative of y with respect to x, regardless of what the independent variable may be.* It is in fact this theorem which justifies the viewpoint that the derivative may be expressed as a ratio of two differentials and these differentials may be handled as if they were members of an ordinary fraction.

The differentials introduced above are finite quantities. As stated above, $\Delta x = dx$ can be chosen arbitrarily, but the expression for dy cannot be deduced until the expression for $D_x y$ has been derived. The result for an anti^d is independent of the magnitude of dx, since $d^{-1} f(x) \, dx = F(x) + C$, whatever be the value of dx. In general, the actual increment Δy does not equal dy, because the latter is the hypothetical increase in y which would have taken place if y had continued to change at the same rate as it did at the point x.

To recover Leibnizian differentials from their finite counterparts introduced above, we will now explore the consequences of regarding Δy and Δx as infinitesimals, the situation considered in defining the derivative. In this context, it is appropriate to call dy the *principal part* of the increment Δy (see pp. 253–254 for a definition of the term). The difference $\Delta y - dy$ may be expressed in the form

$$\Delta y - dy = \underbrace{\left[\frac{\Delta y}{\Delta x} - D_x y \right]}_{\to 0 \text{ as } \Delta x \to 0} \Delta x, \tag{1.20}$$

and one may infer, from the fact that the expression within the square brackets tends to zero as $\Delta x \to 0$, the following theorem:

16

If x is the independent variable and if Δy and dy are regarded as in-finitesimals, the difference $\Delta y - dy$ is an infinitesimal of higher order than Δx.

The chain rule

If y is a composite function—that is, if $y = f(u) = f(g(x))$—the derivative $D_x y$ can be calculated by using the formula

$$D_x y = D_u y \cdot D_x u, \tag{1.21}$$

usually known as the *chain rule* of differentiation. It is custom-ary to call u the intermediate variable, and x the independent variable. In Eq. (1.18), t is the independent variable, and x the in-termediate variable. Hardy remarked that the "proof of this the-orem requires a little care", and appended the following footnote [73, pp. 217–8]: "The proofs in many text-books (and in the first three editions of this book) are inaccurate." A relatively transpar-ent proof is presented in Appendix A.

In a textbook dealing exclusively with d-Calculus, the term *chain rule* is unlikely to be found. When differentiating a com-posite function, one merely identifies the intermediate variable (which we will take to be u here), and writes

$$dy = f'(u)\, du. \tag{1.22}$$

If dy is to be expressed in terms of the independent variable, one simply substitutes $g(x)$ for u and $g'(x)\, dx$ for du. For example, if $y = \sin x^2 = \sin u$,

$$dy = \cos u\, du = \cos x^2 \cdot 2x\, dx. \tag{1.23}$$

One sees from Eq. (1.22) that the differential of a composite function has the same form as it would have if the intermediate variable were the independent variable. In other words, the form of the differential does not depend on whether the argument of the function is an independent variable or the function of an-other variable. This important property of a differential is called the *invariance of the form of the differential* [74–76]. To paraphrase the words of one author [77, p. 72] among many who could be quoted here, Eq. (1.22) shows that taking a differential may be interpreted as differentiation with respect to an arbitrary or un-specified parameter (provided of course that u is differentiable in this parameter).

On dividing Eq. (1.22) by dx, and replacing $f'(u)$ by dy/du, one obtains

$$\frac{dy}{dx} = \frac{dy}{du} \cdot \frac{du}{dx},\qquad(1.24)$$

a mathematical aphorism that is more memorable than its counterpart in D-Calculus, stated in Eq. (1.21).

The importance of remembering the day of the week

In H-Calculus, the meanings of the symbols dy/dx and $\int f(x)\,dx$ depend on which 'notational era or age' one places oneself in: the *age of D-ism* (for short AoD) or the *age of ecumenism* (for short AoE). One should bear in mind that, in the AoD, dy and dx have no separate existence and dy/dx represents not a fraction but the limit of a fraction, much better represented by the symbol $D_x y$; in the AoE, one amalgamates D-Calculus with d-Calculus, which cannot be done without accepting dx and dy as stand-alone symbols, linked by the relation $dy = D_x y\,dx$. Confusion is bound to arise if the reader or the author of a book loses sight of this fact.

ecumenism: the organized attempt to bring about unity among different Christian churches.

The authors of a book on H-Calculus write [78, p. 100]: "The derivative of the function f is the function f' defined by $f'(x) = \lim_{h\to 0}\{[f(x+h) - f(x)]/h\}$ for all x for which this limit exists." Three pages later, the symbol dy/dx makes its appearance [78, p. 103]:

$$\frac{dy}{dx} = \lim_{\Delta x \to 0} \frac{\Delta y}{\Delta x},$$

and the authors continue:

Hence, if $y = f(x)$, we often write

$$\frac{dy}{dx} = f'(x)$$

(The so-called differentials dy and dx are discussed carefully in Chapter 4.) The symbols $f'(x)$ and dy/dx for the derivative of the function $y = f(x)$ are used almost interchangeably in mathematics and its applications, so you need to be familiar with both versions of the notation. You also need to know that dy/dx is a single symbol representing the derivative; it is *not* the quotient of two separate quantities dy and dx.

A new symbol is introduced on p. 114: "A third notation is sometimes used for the derivative $f'(x)$: it is $D_x f(x)$". Differentials are defined on p. 205, and dy/dx promptly becomes the quotient of dx and dy. One might expect that this is the beginning of the AoE, but the authors seem to lose track of their train of thought when they decide to solve a differential equation [78, p. 458]:

> [W]e have the differential equation
>
> $$\frac{dx}{dt} = kx, \tag{3}$$
>
> which serves as the mathematical model for an extraordinarily wide range of natural phenomena. It is easily solved if we "separate the variables" and then integrate:
>
> $$\frac{dx}{x} = k\,dt; \tag{a}$$
>
> $$\int \frac{dx}{x} = \int k\,dt; \tag{b}$$
>
> $$\ln x = kt + C. \tag{c}$$

The letter labels (*a–c*) for the equations are not present in the original text. The authors, who are using H-Calculus, enclose the phrase *separate the variables* within quotation marks because they place themselves in the AoD, where the left hand side of their Eq. (3) stands for an indecomposable symbol, and the dx and dy are supposed to be inseparable. Insertion of a pair of quotation marks, being no substitute for a sound mathematical argument, creates a problem for the student who has been exposed to only H-Calculus. Such a student can integrate an equation of the type

$$D_t x = f(t)$$

in which each side is a finite function (with no $dt, dx, dy, \cdots$), by using the recipe

$$\int (\text{left hand side})\,d\xi = \int (\text{right hand side})\,d\xi,$$

in which the choice of the integration variable ξ is left to the discretion of whoever is carrying out the integration. But a student familiar with the foregoing recipe will find Eq. (*a*) perplexing, since each side here has a symbol (*dx* on the left, *dt* on the right) that is not supposed to have a separate existence in the AoD. Even if the student ignores the anomaly and nonchalantly applies $\int$ to each side of the equation, (s)he might ask: "How come that I am integrating with respect to *x* on one side and with respect to *t* on the other side?".

The truth is that "separation of variables" is permissible only in *d*-Calculus. If we want to avail ourselves of this move, we must fast-forward H-Calculus to the *age of ecumenism* and follow the rules of *d*-Calculus, which requires treating *dx*/*dt* in Eq. (3) as a fraction. Separation of variables, now a legitimate process, leads to Eq. (*a*), each side of which is a differential. To find the corresponding antidifferentials, we apply $\int \equiv d^{-1}$ to both sides, thereby integrating each differential as it stands. The differential on the left is expressed in terms of *x*, but the result will not change if we express it in terms of *t* before applying $\int$ to it (see below).

We still have the option (or the right) of staying in the *age of D-ism* in H-Calculus. Let us go on to see how the problem is to be solved within a regime where no one can put asunder, even by using quotation marks, what the god of *D*-ism has joined together.

Solution using D-Calculus *and* H-Calculus of the AoD

Since differentials are not to be mentioned now, it would forestall confusion if we use $D_t x$ instead of *dx*/*dt*, and start by replacing Eq. (3) with the equivalent derivative equation

$$D_t x = kx. \tag{3^D}$$

We divide both sides of the last equation by *x* (which is permitted, since we are interested only in those situations where $x > 0$) to get the counterparts of Eq. (*a*):

$$\frac{D_t x}{x} = k. \tag{a^*}$$

Each side of Eq. (*a**) is a finite function, and we seek their antiderivatives. Contrary to the previous case, we must now specify a variable with respect to which both sides are to be inte-grated. What should we say (and do) after writing Eq. ($a*$)? We should say "Integrate Eq. ($a*$) *with respect to t*", and go on to write

$$\underbrace{\int_t \frac{1}{x} D_t x = \int_t k}_{\text{Eq. }(b^*)} \quad \text{or} \quad \underbrace{\int \frac{1}{x} D_t x \, dt = \int k \, dt,}_{\text{Eq. }(b')} \qquad (1.25)$$

depending on which species of calculus we are using.

The integrand becomes tractable when we invoke the chain rule:

$$\int_t \frac{D_t x}{x} = \int_t D_x \ln x \cdot D_t x = \int_t D_t \ln x = \ln x \qquad (1.26)$$

Whence follows the result

$$\ln x = kt + C. \qquad (1.27)$$

Back to d-Calculus

As stated above, d-Calculus permits separation of variables without any apologies, and all steps between Eqs. (3) and (c) are perfectly legitimate. However, the passage from Eq. (a) to Eq. (b) may seem odd to those who would like to integrate both sides of Eq. (a) with respect to the same variable. The following argument might appear more convincing to them.

Division of Eq. (3) by x gives

$$\frac{1}{x} \frac{dx}{dt} = k, \qquad (1.28)$$

but this equation, since it has a finite function on each side, can-not be integrated immediately in d-Calculus, where integration is performed on differentials. Accordingly, we first multiply each side with a differential, say $d\zeta$, and then integrate the resulting equation to get

$$\int \frac{1}{x} \frac{dx}{dt} d\zeta = \int k \, d\zeta. \qquad (1.29)$$

All we need do now is to set $dt = d\zeta$, because cancellation of differentials is a legitimate operation in d-Calculus:

$$\int \frac{1}{x}\frac{dx}{dt}\,dt = \int \frac{1}{x}\frac{dx}{dt}\,dt = \int \frac{dx}{x} = \ln x. \qquad (1.30)$$

We see now that the longer route, starting with Eq. (1.28), leads to the same result as that found by separating the variables and integrating the separated form.

The quality of good symbols

The relative ease with which the solution to Eq. (3) emerges if one employs the apparatus of d-Calculus falls in the category to which one may apply the following comment of Whitehead, who was referring to a different example [79, p. 61]:

> This example shows that, by the aid of symbolism, we can make transitions in reasoning almost mechanically by the eye, which otherwise would call into play the higher faculties of the brain.
>
> It is a profoundly erroneous truism, repeated by all copy-books and by eminent people when they are making speeches, that we should cultivate the habit of thinking of what we are doing. The precise opposite is the case. Civilization advances by extending the number of important operations which we can perform without thinking about them. Operations of thought are like cavalry charges in a battle—they are strictly limited in number, they require fresh horses, and must only be made at decisive moments.

I wonder if Whitehead needed a pause for *thinking*, for asking himself if a profoundly erroneous notion could be called a truism, for deliberating whether he should rephrase and write "It is a widespread untruism, repeated by ...". I am sure he was not an advocate of learning (or teaching) mathematics without *ever* thinking about what we are doing, but there is a great deal of down-to-earth pragmatism in his emphatic disapproval of anyone who insists that we should cultivate the habit of thinking about what we are doing even when we carry out tasks that have to performed over and over and over again. Thinking would then get in the way of doing if we followed such a policy.

I happen to share Menger's views, quoted on pp. 237–238, on the consequences of the decision to incorporate the symbols dy/dx and $\int f(x)\,dx$ in H-Calculus. The notation of d-Calculus is a mixed blessing; it allows a beginner to carry out some routine operations without thinking, but it also helps create, and then continually reinforce, the impression that many important rules and theorems—for example, the formulas for the derivatives of a composite function and an inverse function, or the fundamental theorem of calculus—are perfectly obvious results. In *Macbeth* (Act 2, Scence 3), the porter, talking about drinking, tells Macduff: "Lechery, sir, it provokes, and unprovokes; it provokes the desire, but it takes away the performance". For the teaching of calculus through an (almost) exclusive use of the Leibnizian notation, one may say: It promotes the performance but takes away the understanding.

As for the part played by routine maipulations in advancing civilization, I am inclined to agree with Medawar, who pointed out that Whitehead had stated only a semi-truism [80, pp. 137–8]:

> This is a most important and arresting half-truth—so compellingly true, or half true, that one wonders how anyone could ever have held a contrary opinion. But it is true only of *learned* activity. No matter what the activity may be—learning the multiplication table, or how to drive a car, to speak intelligibly, or to sew—learning is a process of thinking and deliberation and trial and decision, but the *state of having learned* is the state in which one need think no longer. Paradoxically enough, learning is learning not to think about operations that once needed to be thought about; we do in a sense strive to make learning "instinctive", i.e. to give learned behaviour the readiness and aptness and accomplishment which are characteristic of instinctive behaviour. But that is only half the story. The other half of the half truth [of the whole truth?] is that civilization also advances by a process which is the very converse of that which Whitehead described: by learning to think about, adjust, subdue and redirect activities which are thoughtless to begin with because they are instinctive. Civilization also advances by bringing instinctive activities within the domain of rational thought, by making them reasonable, proper and co-operative. Learning, therefore, is a twofold process: we learn to make the processes of deliberate thought "instinctive" and automatic, and we learn to make automatic and instinctive processes the subject of discriminating thought.

Chapter 2

Integration: Part I

IN the context of calculus, the term *integration* is applied to two closely entwined but distinct operations. One of these, named definite integration, can be traced all the way back to geometers like Archimedes; the other, called indefinite integration, emerged as an answer to the quest for an operation that is the inverse of differentiation, just as division is the inverse of multiplication. The rest of this section reproduces the first two paragraphs from the chapter on integration in Hobson's treatise [81, p. 337], mainly because the phrase *fundamental theorem of integral calculus* appears to have been introduced here to the English-reading world.

"The fundamental operation of the calculus, known as integration, regarded from one point of view consists essentially in the determination of the limit of the sum of a finite series of numbers, as the number of terms of the series is indefinitely increased, whilst the numerically greatest of the individual terms of the series approaches the limit zero. The laws which regulate the specification of the terms of the series must be supposed, in any given instance, to be assigned, and to be of such a character that the limit in question exists. It is in this form that the problem of integration naturally presents itself in ordinary problems of a geometrical character, such as the determination of lengths, areas, volumes, &c. The method of integration, so regarded, has its origin in the method of exhaustions employed by the Greek geometers, and was developed later in forms of which the exactitude depended at various epochs upon the stage which the development of Analysis in general had reached. In the hands of Cauchy, Dirichlet, and Riemann the definition of the definite integral attained to the exact arithmetic form in which it is employed in modern analysis; and in fact the definition given by Riemann, which is now held to be fundamental in the calculus, leaves nothing to be desired as regards precision. Riemann gave not only a precise definition, but also a necessary and sufficient condition, for the existence of the definite integral. Although a more general definition of integration has recently been developed by Lebesgue, in accordance with which classes of functions are integrable, which are not so in accordance with the definition of Riemann, the latter is the definition which lies at the base of almost all the developments of the theory of integration that have been made during the last half century [1850–1900] ...

"Integration has also usually been regarded as the operation inverse to that of differentiation; and the *fundamental theorem of the Integral Calculus* formulates the relation of this mode of regarding integration with the one referred to above. Many important investigations are concerned with the relation between these two modes of regarding integration, with the establishment of the fundamental theorem, and with an examination of the limitations to which it is subject." (Emphasis added.)

Scope of the chapter

In this chapter, the term integration will be interpreted as the converse of differentiation. Regardless of which variety of calculus (or what symbol) one uses, integration amounts to looking for a function which will differentiate into the given integrand. One is required to demonstrate that the given integrand, say $f(x)$, can be equated to $D_x F(x)$, or equivalently, that $f(x)\,dx$ can be equated to $dF(x)$). Among the devices used for achieving this end, we will consider only that known as the *method of substitution*, and we will now look into to the basis of this method.

Change of variable in integration: some examples

We are familiar with the fact that if

$$y = f(u) \quad \text{and} \quad u = g(x),\tag{2.1a}$$

then we can write

$$D_x f = D_u f \cdot D_x u = f'(u) \cdot g'(x).\tag{2.1b}$$

In the intermediate step of an actual calculation, the first derivative, namely $f'(u)$, is expressed in terms of u, and is transformed in the final step to a function of x by replacing u with $g(x)$. This remark is illustrated in the following example.

Let $Y = \sin x^2$ and choose $u = g(x) = x^2$, so that $Y = \sin u$:

$$D_x Y = f'(u) \cdot g'(x)\tag{2.2a}$$
$$= \cos u \cdot 2x\tag{2.2b}$$
$$= \cos x^2 \cdot 2x\tag{2.2c}$$

We will now find the x-anti^D of

$$y = 2x \cos x^2$$

by going in the opposite direction.

Let Y be the required x-anti^D, so that $D_x Y = y$. Choose $u = x^2$, so that $D_x u = 2x$, and invoke the chain rule:

$$y = D_x Y = D_u Y \cdot D_x u,\tag{2.3a}$$
$$\text{or} \quad y = D_u Y \cdot 2x\tag{2.3b}$$
$$\therefore D_u Y = \frac{y}{2x} = \frac{2x \cos x^2}{2x} = \cos u.\tag{2.3c}$$

We see that our substitution has reduced the original problem of finding $D_x^{-1}(2x \cos x^2)$ to the simpler task of finding $D_u^{-1}(\cos u)$.

Since the aim of the substitution is to arrive at an expression for $D_u Y$, it is better to begin by writing the chain rule in a different form:

$$D_u Y = D_x Y \cdot D_u x = y \cdot D_u x\tag{2.4a}$$
$$= 2x \cos x^2 \cdot \frac{1}{2x} = \cos x^2 = \cos u\tag{2.4b}$$

In d-Calculus, one would begin by writing the differential version of Eq. (2.4a), and proceed as before:

$$dY = y\,dx = y\frac{dx}{du}\,du \tag{2.5a}$$

$$= 2x\cos u\frac{1}{2x}\,du = \cos u\,du. \tag{2.5b}$$

Note that Eq. (2.5a) can be interpreted as follows: For changing the variable of integration from x to u, use the formula

$$dx = \frac{dx}{du}\,du. \tag{2.6}$$

The upshot of Eqs. (2.4a), (2.5a), and (2.6) may be stated as follows: *in all three species of calculus, a change of variable from x to u involves multiplying the integrand by $D_u x = dx/du$.*

It is not always necessary to use the above recipe, because the notation of d-Calculus permits, as shown below, a shortcut that bypasses Eq. (2.6):

$$dY = y\,dx = 2x\cos x^2\,dx = \cos x^2(2x\,dx) = \cos u\,du.$$

We will close this section by finding the x-anti^D of the function

$$y = \sqrt{(1 - x^2)}. \tag{2.7}$$

We are required to find a function, say Y, whose x-derivative is $y = \sqrt{(1 - x^2)}$. As always, there is more than one way of finding Y, but here we would like to avail ourselves of a substitution that turns y into a function $y = f(u)$ of a new variable, say $u = g(x)$, where g is some trigonometric function chosen with a view to making it easy for us to primitate $f(u)$

We will use the substitution $u = \arcsin x$, so that

$$x = \sin u, \quad D_u x = \cos u, \quad y = \cos u, \quad \text{and}$$

$$D_u Y = D_x Y \cdot D_u x = y \cdot D_u x = \cos^2 u. \tag{2.8}$$

It is not too big a challenge to primitate $\cos^2 u$ (the meaning of *primitate* is explained on p. xxv), and convince oneself that

$$D_u Y = \cos^2 u, \quad \text{if} \quad Y = \tfrac{1}{2}\left[u + \tfrac{1}{2}\sin 2u\right] \tag{2.9}$$

All that remains now is to express u and $\sin 2u$ in terms of x, and one sees immediately that

$$Y = \tfrac{1}{2}\big[\arcsin x + x\sqrt{(1-x^2)}\big] \qquad (2.10)$$

and, since $D_x \arcsin x = 1/\sqrt{(1-x^2)}$,

$$D_x Y = \sqrt{(1-x^2)}, \quad \text{as expected.} \qquad (2.11)$$

It is natural to ask whether a different substitute, namely $v = \arccos x$, would lead to the same result. With this substitution, we have

$$x = \cos v, \quad D_v x = -\sin v, \quad y = \sin v, \quad \text{and}$$

$$D_v Y = D_x Y \cdot D_v x = y \cdot D_v x = -\sin^2 v. \qquad (2.12)$$

$$\therefore Y = D_v^{-1}(-\sin^2 v) = \tfrac{1}{2}\big[-v + \tfrac{1}{2}\sin 2v\big]. \qquad (2.13)$$

Since $v = \arccos x = \tfrac{1}{2}\pi - \arcsin x$, we are now led to the following expressions

$$Y = \tfrac{1}{2}\big[-\arccos x + x\sqrt{(1-x^2)}\big] \qquad (2.14)$$

$$= \tfrac{1}{4}\pi + \tfrac{1}{2}\big[\arcsin x + x\sqrt{(1-x^2)}\big]. \qquad (2.15)$$

Change of variable in integration: a formal summary

Little can be added to the foregoing account of the simplification resulting from a judicious change of the independent variable, but, since we have used only the symbols d and D, it does not seem amiss to summarize this account in terms of formal relations involving the symbol of integration in each of the three species of calculus.

Substitution in d-Calculus

We will consider the integral

$$y = \int \underbrace{f(x)\,dx}_{\text{integrand}} \qquad (2.16)$$

and suppose that the independent variable is to be changed from x to u, with $x = g^{-1}(u) = \phi(u)$. Then

$$y = \int f(x)\,dx = \int f(x)\frac{dx}{du}\,du \qquad (2.17)$$

$$= \int f[\phi(u)]\frac{d\phi}{du}\,du. \qquad (2.18)$$

Substitution in H-Calculus

In H-Calculus the dx in the integral

$$y = \int \underbrace{f(x)}_{\text{integrand}}\,dx \qquad (2.19)$$

is only a device for indicating the integration variable. With the Leibnizian shortcut used in Eq. (2.6), viz. $dx = (dx/du)\,du$, now blocked, a follower of H-calculus must make explicit use of the chain rule to find dy/du, and integrate it over u to deduce an expression for y in terms of u. The route is shown below:

$$D_u y = D_x y \cdot D_u x = f(x)D_u x \qquad (2.20)$$

Integration with respect to u now gives

$$y = \int \underbrace{f(x)\,D_u x}_{u\text{-integrand}}\,du \qquad (2.21)$$

$$= \int f[\phi(u)]D_u[\phi(u)]\,du \qquad (2.22)$$

The rules of H-Calculus demand that du be appedded to the right hand side of Eq. (2.21) to display the variable of integration. The truth is that students simply skip the parentheses and arrive at the correct result because $D_x y = dy/dx$.

Substitution in D-Calculus

The argument is precisely the same as in the preceding derivation, but the notation is different. One now writes

$$y = \int_x f(x), \qquad (2.23)$$

so that

$$D_x y = f(x). \qquad (2.24)$$

When integration is to be performed with respect to a new variable $u = g(x)$, with $x = g^{-1}(u) = \phi(u)$, we write Eq.(2.20) as

$$D_u y = D_x y \cdot D_u x = f(x) \cdot D_u x, \qquad (2.25)$$

and integrate both sides with respect to u to get

$$y = \int_u f(x) D_u x \qquad (2.26)$$

$$= \int_u f[\phi(u)] D_u[\phi(u)]. \qquad (2.27)$$

The prescriptions given in Eqs. (2.17), (2.21), (2.26) and (2.27) are as close as they ought to be, but the differential route to a particular result is less demanding; in finding the anti^d of $dy = f(x)\,dx$, the presence of the differential dx reduces a change of variable to the trivial task of tinkering with fractions. If the person (student or teacher) who is doing the tinkering understands the rationale behind the tinkering, and is able to explain, when asked to do so (in an examination or in the course of teaching), how one may go from the left hand side of Eq. (2.21) to the right hand side without using differentials, there is really no cause for complaint or concern.

We established the identity

$$\int_x f(x) = \int_u f[\phi(u)] D_u x, \quad \text{if} \quad x = \phi(u) \qquad (2.28)$$

by taking the chain rule as our starting point. We will now go in the reverse direction and prove the identity by differentiating both sides with respect to x and showing that the resulting expressions are identical. The x-derivative of the right hand side is simply $f(x)$, because $D_x \int_x = 1$. The x-derivative of the u-integral is calculated below:

$$D_x \int_u f[\phi(u)] D_u x = \underbrace{\left[D_u \int_u f[\phi(u)] D_u x \right]}_{=1} \cdot D_x u \qquad (2.29)$$

$$= f[\phi(u)] D_u x \cdot D_x u = f[\phi(u)], \qquad (2.30)$$

which proves the identity.

30

Bringing D-Calculus and d-Calculus still closer

Let us note that integration of the differential equation

$$dy = y'(x) \cdot dx = f(x) \cdot dx, \tag{2.31}$$

leads to Eq. (2.19):

$$y = \int f(x)\,dx. \tag{2.19}$$

Here the integration variable is taken to be x, but if one wishes to change to some other variable, say u, one simply goes on to write:

$$y = \int f(x)\,dx = \int f(x)\,\frac{dx}{du} \cdot dx. \tag{2.32}$$

We now ask if it is possible to deduce analogous relations in which d is replaced by D.

Since the ultimate variable in Eq. (2.25) can be any variable (so long as x can be differentiated with respect to it), we now drop the symbol u and replace Eq. (2.25) with that shown below:

$$Dy = D_x y \cdot Dx = f(x) \cdot Dx, \tag{2.33}$$

which is the counterpart of Eq. (2.31). We note in passing

$$\frac{Dy}{Dx} = D_x y = \frac{dy}{dx}. \tag{2.34}$$

We now integrate Eq. (2.33) with the understanding that both D and $\int$ represent mutually inverse operations (of differentiation and integration, respectively) with respect to the same arbitrary and unspecified variable, and state the result of integration as

$$y = \int Dy = \int f(x) \cdot Dx, \tag{2.35}$$

which is the D-analog of Eq. (2.19).

If one knows how to find an x-primitive of $f(x)$—that is, a function whose x-derivative equals $f(x)$—one would identify x with the arbitrary variable; though $Dx = 1$ now, its presence serves as a reminder of the variable of integration. If a substitution, say $x = \phi(u)$, promises an easier path, the change is implemented as follows:

$$\int f(x) \cdot Dx = \int f[\phi(u)] \cdot \frac{Dx}{Du} \cdot Du = \int f[\phi(u)]D_u x \cdot Du. \tag{2.36}$$

If a u-primitive of $f[\phi(u)]D_u x$ can be found, one would choose u as the arbitrary variable, which makes $Du = 1$, of course. The result will not change if u is later identified as a function of some other variable.

The foregoing is a paraphrase of John West's approach (see Chapter 14). He integrated, without explaining why, functions of the form $f(x)Dx$ instead of $f(x)$. Another reason for presenting this material here is to draw attention to the point of view according to which differentials may be regarded as derivatives with respect to an arbitrary parameter [82, 83].

A few clerical details

Of the ten questions raised in the Preamble, only the last remains to be answered. The person who posted Q-10 began by stating the formula for a change of variables in a 2-dimensional integral, and finished with the following comments: "It seems that the formula for 2D (and higher dimensions) are extensions of the 1D version. If so, then the 1D version 'should' also require absolute value. Can anyone explain why, except for 1D, do all higher versions require the absolute value of Jacobians?" None of the respondents hit the nail on the head, but everyone seemed to agree (as far as I could make out) that the one-variable integration is an exceptional case. This conclusion, if well founded, would drive every student into agreeing with another disappointed and irate undergraduate who thought [84, p. 67]: "My mathematical tutors had never shown me any reason to suppose the Calculus anything but a tissue of fallacies". The procedure for integration by substitution (in the one-variable case) will be summarized in this section, and it will be shown later that a logical extension of the same procedure is capable of dealing with integrals involving two or more variables.

Some new notation

As a preliminary to what will be said in this section and the rest of the book, we need some new notation for representing differences of the type shown below:

$$\left[U(x) \right]_a^b = U(x) \Big|_a^b = U(b) - U(a). \tag{2.37}$$

Here x is a dummy variable, which is another way of saying that x can be replaced by any other symbol (apart from a and b). We will assume that $a < b$.

If $F(x)$ is a function whose x-derivative is $f(x)$, we can write

$$\left[\int_x f(x)\right]_{x_1}^{x_2} = \left[\int f(x)\,dx\right]_{x_1}^{x_2} = F(x_2) - F(x_1) \qquad (2.38)$$

In writing Eq. (2.38), both notations for the anti^D have been used, but it is rather inconvenient to keep doing this. I will therefore use only the suffix notation in the rest of the section; a reader who finds it easier to handle the conventional notation can drop the suffix ζ and replace it with $d\zeta$ ($\zeta = x$ or u).

If $u = g(x)$, $x = g^{-1}(u) = \phi(u)$, $F[\phi(u)] = \Phi(u)$, $u_1 = g(x_1)$, $u_2 = g(x_2)$, we can effect a change of variable by using Eq. (2.28), and writing

$$\left[\int_x f(x)\right]_{x_1}^{x_2} = \left[\int_u f[\phi(u)]D_u x\right]_{u_1}^{u_2} = \Phi(u_2) - \Phi(u_1). \qquad (2.39)$$

Since it is customary to place the smaller value at the lower index, we must now distinguish, if we want to follow this custom, between two cases: (i) u is monotonically increasing so that $D_u x > 0$ for all x and $u_1 < u_2$, and (ii) u is monotonically decreasing so that $D_u x < 0$ for all x and $u_1 > u_2$. Let u_- and u_+ denote the smaller and the larger (respectively) of u_1 and u_2; symbolically, $u_- = \min(u_1, u_2)$ and $u_+ = \max(u_1, u_2)$.

We will now state the formulas for the change of variable in terms of $|D_u x|$ and $u_\pm$. In case (i), $D_u x = |D_u x|$, $u_1 = u_-$, $u_2 = u_+$, and

$$\left[\int_u f[\phi(u)] \cdot D_u x\right]_{u_1}^{u_2} = \left[\int f(\phi(u)) \cdot |D_u x|\right]_{u_-}^{u_+} \qquad (2.40)$$

In case (ii), $D_u x = -|D_u x|$, $u_1 = u_+$, $u_2 = u_-$, and

$$\left[\int_u f[\phi(u)] \cdot D_u x\right]_{u_1}^{u_2} = -\left[\int f[\phi(u)]|D_u x|\right]_{u_+}^{u_-} \qquad (2.41)$$

$$= \left[\int_u f[\phi(u)] \cdot |D_u x|\right]_{u_-}^{u_+} \qquad (2.42)$$

For the sake of condensing all the relevant results into a single equation with a uniform notation, we will replace the lower limit x_1 by x_- and the upper limit x_2 by x_+.

$$\left[\int_x f(x)\right]_{x_-}^{x_+} = \left[\int_u f[\phi(u)] \cdot |D_u x|\right]_{u_-}^{u_+} \tag{2.43}$$

Most beginners would use Eq. (2.39) rather than Eq. (2.43), and anyone who solves more than one or two integration problems involving substitution soon discovers that whenever $D_u x$ is negative, $u_1 > u_2$, and realizes that the negative sign can be dropped if one takes u_2 as the lower limit and u_1 as the upper limit. It should also be added that the argument involving the absolute value of $D_u x$ might not have been added above if the author had not come across Q-10, where the questioner asks: Why absolute values of Jacobians in change of variables for multiple integrals but not single integrals? Introduction of the absolute value is not a necessity in either case, but it turns integration by substitution into a routine process, because the lower limit in each integral is always smaller than the corresponding upper limit.

A symbol with more than one meaning

Several examples will be presented in this book to convince the reader that the symbol $\int(\ldots)\,dx$ is used for two different (albeit closely related) concepts. I have come across only one author who informs his readers that the standard notation for the indefinite integral is made to serve double duty [85, p. 183].

> Thus the symbol $\int f(x)\,dx$ denotes the collection of all antiderivatives of the function $f(x)$. But sometimes we shall understand it to be any element of that collection, i.e. any one of the antiderivatives.

Several illustrations of the above mentioned mathematical doublespeak are listed in Appendix C. Here a single example is considered sufficient.

One example

I reproduce below the opening remarks from a book on differential equations [86], jointly written by Edwards and Penney

34

(hereafter E&P); in what follows, Eq. (n) of the original text has been numbered as Eq. (p:n), where p specifies the page number.

The first-order equation $dy/dx = f(x,y)$ takes an especially simple form if the right-hand-side function f does not actually involve the dependent variable y, so

$$\frac{dy}{dx} = f(x). \tag{10:1}$$

In this special case we need only integrate both sides of Eq. (1) to obtain

$$y(x) = \int f(x)\,dx + C. \tag{10:2}$$

This is a general solution of Eq. (1), meaning that it involves an arbitrary constant C, and for every choice of C it is a solution of the differential equation in (1). If $G(x)$ is a particular anti^D of f—that is, if $G'(x) \equiv f(x)$—then

$$y(x) = G(x) + C. \tag{10:3}$$

On comparing the right hand sides of Eqs. (10:2) and (10:3), one sees that the integral

$$\int f(x)\,dx$$

stands for a particular anti^D, not the most general anti^D. However, on the next page of the same book one finds that the *general* solution of the differential equation

$$\frac{dy}{dx} = 2x + 3$$

is stated to be

$$y(x) = \int (2x+3)\,dx = x^2 + 3x + C.$$

In the last equation $\int (\cdots)\,dx$ is interpreted, in accordance with the standard practice, as the most general primitive of the function within the parentheses.

Using the same symbol for two different purposes is acceptable if the reader has acquired the requisite maturity, but I will avoid ambiguity by using two different symbols (or a caution that I am merely quoting other authors). The following equation spells out the distinction:

$$\int f(x)\,dx = \underbrace{\int^\dagger f(x)\,dx}_{\text{variable part}} + C, \tag{2.44}$$

The first term on the right hand side of Eq. (2.44) is an anti^d of $f(x)\,dx$ to which no constant has been *purposely* added. In D-Calculus, the contents of the above equation would be written as

$$\int_x f(x) = \int_x^\dagger f(x) + C, \tag{2.45}$$

If one is asked to choose (or coin) an apt name for the integral on the right hand sides of Eqs. (2.44) and (2.45), one might be tempted to call it the *simplest primitive*. This is one of those occasions, where one would be well advised to count (at least) to two, if not to the proverbial ten, and consider some concrete examples, before proposing a name.

Let us note, for example, that if we are asked to find the anti^d of $(\alpha + \beta x)^2\,dx$, we have two choices. We may expand $(\alpha + \beta x)^2$, and integrate term by term:

$$\int^\dagger (\alpha + \beta x)^2\,dx = \int^\dagger (\alpha^2 + 2\alpha\beta x + \beta^2 x^2)\,dx$$

$$= \alpha^2 \int^\dagger dx + 2\alpha\beta \int^\dagger x\,dx + \beta^2 \int^\dagger x^2 dx$$

$$= \alpha^2 x + \alpha\beta x^2 + \tfrac{1}{3}\beta^2 x^3 = Y_1, \text{ say.} \tag{2.46}$$

Alternatively, we may decide to use $\alpha + \beta x$ as the integration variable, so that

$$Y_2 \equiv \int^\dagger (\alpha + \beta x)^2\,dx = \frac{1}{\beta} \int^\dagger (\alpha + \beta x)^2 d(\alpha + \beta x),$$

$$= \frac{1}{3\beta}(\alpha + \beta x)^3$$

$$= (\alpha^3/3\beta) + \alpha^2 x + \alpha\beta x^2 + \tfrac{1}{3}\beta^2 x^3$$

$$= (\alpha^3/3\beta) + Y_1. \tag{2.47}$$

We see that Y_1 and Y_2 differ by a non-arbitrary constant. In this particular example, and in other cases of this type, it would be natural to drop the non-arbitrary constant (by adding $-\alpha^3/3\beta$), and it would not be unreasonable to call Y_1 the simplest anti^D of $(\alpha + \beta x)^2$.

However, it is not always so easy to spot the simplest anti^D; different routes for finding integrals may lead to two (or more primitives) *none* of which contains a non-arbitrary additive constant. Depending on the method used for integrating $f(x) = \sin x \cos x$, one is led to three primitives:

$$F_1(x) = \tfrac{1}{2}\sin^2 x, \tag{2.48a}$$

$$F_2(x) = -\tfrac{1}{2}\cos^2 x = F_1(x) - \tfrac{1}{2}, \tag{2.48b}$$

$$F_3(x) = -\tfrac{1}{4}\cos 2x = F_1(x) - \tfrac{1}{4}. \tag{2.48c}$$

All three have equal claim to being called the simplest primitive, for none of them has an additive constant, and all three satisfy the criterion

$$D_x F_j = f(x), \qquad (j = 1,2,3). \tag{2.49}$$

Let us go back to example considered on p. 34, where integration was regarded as a task that entailed solving the equation

$$\frac{dy}{dx} = f(x). \tag{2.50}$$

In accordance with the standard terminology, any function that satisfies the above differential equation will be called a *particular integral* of the above equation. If we choose, to take a concrete example, $f(x) = (\alpha + \beta x)^2$, then Y_1 and Y_2 are both particular integrals (or particular antiderivatives or particular antidifferentials), and in a practical application, we will need only one particular integral.

No *arbitrary* constants please, we aren't mathematicians

Charles Darwin considered the appendix to be a vestigial organ, no longer useful, just a relic derived from a larger structure in a common ancestor shared by humans and other apes. This view was aptly summarized in a book entitled *Evolution and Disease* [87, p. 65]: " In man the vermiform appendix is a typical example of a functionless part, and, like an idle person in a community, is not infrequently a source of considerable danger and suffering, and is responsible for a number of deaths annually." Recently, it has been suggested that the appendix may subserve some useful purpose after all. The arbitrary constant seems to me to be an analog of the vermiform appendix.

An arbitrary constant, though it has some pedagogical value (see below), is of no particular use to those who wish to *apply* the technique of integration to a concrete problem.

We will add a constant, say A, to a primitive $F(x)$ only when we are looking for a function, say $G(x)$, that is supposed to satisfy, in addition to the principal requirement

$$D_x G(x) = D_x F(x) = f(x), \qquad (2.51)$$

a subsidiary condition that can be stated as

$$G(x) = A \qquad \text{when } x = a, \qquad (2.52)$$

where both a and A are known numbers.

For instance, $Y_1 = \alpha^2 x + \alpha\beta x^2 + \frac{1}{3}\beta^2 x^3$ is a primitive of $f(x) = (\alpha + \beta x)^2$, but if we are seeking a primitive of $(2 + 3x)^2$ which must have the value of 2 when $x = \frac{1}{3}$, we will be obliged to take $Y = Y_1 + A$ (with $\alpha = 2$, $\beta = 3$), and adjust the value of A by imposing the condition

$$2 = Y\big|_{x=1/3} = Y_1\big|_{x=1/3} + A, \quad \Longrightarrow A = 2 - \tfrac{19}{9} = -\tfrac{1}{9}, \quad (2.53)$$

which means that the particular integral we seek is

$$Y(x) = 4x + 6x^2 + 3x^3 - \tfrac{1}{9}. \qquad (2.54)$$

Let us note in passing that a more elegant way of finding a primitive that is required to fulfil an additional condition will be presented in the next chapter.

When we decide to add a constant with the specific purpose of satisfying a subsidiary condition, we will call it an *adjustable constant* or a *constant yet to be determined*. I owe the latter phrase to Airy, who wrote [88, p. 2]:

> ...a constant of that class described (perhaps improperly) by the term "arbitrary," but which really means "not yet determined, but enabling us by proper determination of its value so to fix the value of x corresponding to a given value of y that we can adjust the

solution to some specific condition."

The reader is requested to observe that instead of the term "arbitrary constant," we shall always use the term "undetermined constant."

What De Morgan wrote about this issue in 1836 is worth recalling [89, p. 105]:

It is usual to omit the constant, as an attendant of the integral sign so well known that it is unnecessary except where we are actually applying the integral calculus, and may be dispensed with when we are merely ascertaining integral forms.

It might have been usual in those days to omit the constant of integration, but the opposite seems to be true now. The addition or omission of an (the?) arbitrary constant seems to have become a shibboleth, used by most teachers as an excuse for deducting a mark [90, 91]. For my part, I am happy to follow De Morgan.

An arbitrary constant not *completely* useless

One of the first integration formulas a student encounters is that shown below

$$\int x^n \, dx = \frac{x^{n+1}}{n+1}, \qquad (n \neq 1). \tag{2.55}$$

That this formula becomes inapplicable when the denominator vanishes is a source of some disappointment to beginners, but we can dodge the restriction by adding an arbitrary constant to the right hand side, and assigning a suitable value to the constant, where by suitable we mean a choice that turns the right hand side into a vanishing fraction (for $n = -1$), whose limiting value will provide the required primitive. Accordingly, we proceed as follows:

$$\int x^n \, dx = \lim_{n \to -1} \frac{x^{n+1} - a^{n+1}}{n+1} = \ln\left(\frac{x}{a}\right), \tag{2.56}$$

and find that the limiting value of the vanishing fraction comes out to be $\ln(x/a)$, or simply $\ln x$, if we set $a = 1$.

Chapter 3

Integration: Part II

Almost all college textbooks have differentiation before integration, but this is precisely the worst way to teach students. Let us begin with integration and calculate areas and volumes by summing series treated in the previous section of the course. This is the only way to drive home that an integral is the limit of a sequence of sums. When this is well established, we turn to the derivatives of polynomials, note the connection with integration, and prove the Fundamental Theorem.

C. B. Allendoerfer [92]

Courant's two-volume calculus book was never widely used because, among other reasons, integration is presented before differentiation. Many people believe that it is easier and sounder pedagogically to teach integration first, but such beliefs have no import, since all calculus books must be similar if the authors expect wide adoptions, and all the other calculus books cover differentiation first.

M. H. Protter [43]

E VEN if the calculus secretariat is as conservative as claimed by Protter, *we* should keep an open mind, and decide on our own if the teaching/learning of calculus would benefit from a reversal of the integration-after-differentiation policy adopted in standard textbooks. I heartily recommend Courant's book [93], a pedagogical masterpiece now available in the public domain.

The definite integral

Take two points A and B on the x-axis, and let a and b be their x-coordinates, respectively (with $a < b$). Let us divide the interval $[a, b]$ into $n + 1$ subintervals, $[a, x_1], [x_1, x_2], [x_2, x_3], \ldots [x_n, b]$. For notational convenience, we will often set $x_0 = a$ and $x_{n+1} = b$. Let us begin by defining the sum

$$S_n = (a - x_1)f(a) + (x_1 - x_2)f(x_1) + \cdots + (b - x_n)f(x_n), \quad (3.1)$$

each term of which may be viewed as the area of an elementary rectangle of width $\Delta x_k = x_{k+1} - x_k$ and height $f(x_k)$.

The definite integral of $f(x) \, dx$ from a to b is defined as the limiting value (as $n \to \infty$) of S_n:

$$\int_a^b f(x) \, dx = \lim_{n \to \infty} S_n. \quad (3.2)$$

In the infinite sum above (which runs from $x = a$ to $x = b$), each term is a differential that may be viewed as an infinitesimal elementary rectangle); accordingly, the definite integral itself may be interpreted as the area of the plane figure bounded by the curve $y = f(x)$, the x-axis and the ordinates at $x = a$ and $x = b$.

In some particularly favourable cases, the sum on the right hand side of Eq. (3.2) can be transformed, using algebraic manipulations, to an analytical expression involving elementary functions; if the problem cannot be solved algebraically, we can resort to numerical methods for finding S_n for a sufficiently large value of n.

The definite integral on the left-hand side of Eq. (3.2) was defined under the supposition that $a < b$. We would now extent the concept to cover the case when the two endpoints (limits) coincide, or when the lower limit happens to be larger than the upper limit.

Principal properties of the definite integral

1. By definition, we assume that

$$\int_a^a f(x) \, dx = 0. \quad (3.3)$$

Also by definition, we assume that

$$\int_a^b f(x)\,dx = -\int_b^a f(x)\,dx = 0. \tag{3.4}$$

2. Whatever the numbers a, b, c, the following additive property

$$\int_a^b f(x)\,dx = \int_a^c f(x)\,dx + \int_c^b f(x)\,dx \tag{3.5}$$

holds true, provided (of course) that the three integrals exist.

3.
$$\int_a^b kf(x)\,dx = k\int_a^b f(x)\,dx \tag{3.6}$$

4.
$$\int_a^b [f(x) \pm g(x)]\,dx = \int_a^b f(x)\,dx \pm \int_a^b g(x)\,dx \tag{3.7}$$

The Newton-Leibniz formula

A result of great importance, named the Newton-Leibniz formula (or theorem) in some Russian books, will be presented next. Using the notation

$$\left[U(x)\right]_a^b = U(x)\Big|_a^b = U(b) - U(a), \tag{2.37}$$

the Newton-Leibniz formula can be stated as follows:

$$\int_a^b f(x)\,dx = \left[\int f(x)\,dx\right]_a^b. \tag{3.8}$$

The left-hand side of Eq. (3.8) is an infinite sum, whereas the right hand side,

$$\left[\int f(x)\,dx\right]_a^b = \left[F(x) + C\right]_a^b = F(x)\Big|_a^b = F(b) - F(a), \tag{3.9}$$

is a difference between two values of a function, the function in question being *any* primitive of the integrand of $\int f(x)\,dx$.

A notation that uses the integral sign only on the left hand side of Eq. (3.8) might make it easier to advertise the fact that Eq. (3.8) connects two numbers whose interrelationship is far from obvious, one defined as an area and the other as the difference between two values of a function. The symbols d^{-1} and D_x^{-1} (introduced in Chapter1) can be used to state the Newton-Leibniz formula [Eq. (3.8)] as

$$\int_a^b f(x)dx = \left[d^{-1}f(x)\,dx\right]_a^b = \left[D_x^{-1}f(x)\right]_a^b. \qquad (3.10)$$

It will not be amiss to reproduce here a footnote in which Piskunov explains that the eponyms did not actually state the above result [94, p. 405]: "It is necessary to point out that the name of formula (2) [which is stated in the form $\int_a^b f(x)dx = F(b) - F(a)$] is not exact, since neither Newton nor Leibniz had any such formula in the exact meaning of the word. The important thing, however, is that namely Leibniz and Newton were the first to establish a relationship between integration and differentiation, thus making possible the rule for evaluating indefinite integrals."

Integrals with a variable endpoint

When we refer to the integral on the left-hand side of Eq. (3.10) as a definite integral, it is implicitly assumed that definite values (such as 0, or 1 or π) have been assigned to the endpoints a and b (the upper and lower limits). But the symbols a and b are to be regarded as parameters, and we are free to vary one limit while keeping the other fixed.

If we relabel the upper limit as x, and change the integration variable to some other symbol, say ξ, we find that the integral

$$\int_a^x f(\xi)\,d\xi = \Phi(x) \qquad (3.11)$$

is now a function of x. By virtue of Eq. (3.3), we have $\Phi(a) = 0$, and two such integrals with different lower limits will differ, by virtue of Eq. (3.5) only by a constant.

It goes without saying that we are entitled to define an indefinite integral to be a function of the lower limit, and write

$$\int_x^b f(\xi)\,d\xi = \Psi(x). \qquad (3.12)$$

As before, that two such integrals with different upper limits will differ only by a constant.

Our next task is to prove that $\Phi(x)$ is an anti^D of $f(x)$; in other words, we are going to show that

$$\frac{d\Phi}{dx} = f(x). \tag{3.13}$$

How can we verify Eq. (3.13). The only way is to differentiate Eq. (3.11) with respect to x, and see whether the answer turns out to be the same as the x-derivative of the right hand side, namely $f(x)$. To see that this is indeed true, we use Eq. (3.5) and write

$$\underbrace{\int_a^{x+\Delta x} f(\xi)\, d\xi}_{=\Phi(x+\Delta x)} = \underbrace{\int_a^x f(\xi)\, d\xi}_{=\Phi(x)} + \underbrace{\int_x^{\Delta x} f(\xi)\, d\xi}_{=K(x,\Delta x)}, \tag{3.14}$$

The required derivative comes out to be:

$$\frac{d\Phi}{dx} = \lim_{\Delta x \to 0} \frac{\Phi(x+\Delta x) - \Phi(x)}{\Delta x} = \lim_{\Delta x \to 0} \frac{K(x,\Delta x)}{\Delta x} \tag{3.15}$$

Since the second integral on the right hand side of Eq. (3.14) extends from $\xi = x$ to $\xi = x + \Delta x$, and we are going to impose the limit $\Delta x \to 0$, we can replace $f(\xi)$ by $f(x)$, and pull it out of the integral to write

$$K(x,\Delta x) = \int_x^{x+\Delta x} f(\xi)\, d\xi = f(x) \int_x^{x+\Delta x} d\xi = f(x)\Delta x, \tag{3.16}$$

and we see that

$$\frac{d\Phi(x)}{dx} = \lim_{\Delta x \to 0} \frac{K(x,\Delta x)}{\Delta x} = \lim_{\Delta x \to 0} \frac{f(x)\Delta x}{\Delta x} = f(x). \tag{3.17}$$

Though the steps taken in Eq. (3.16) will not be acceptable to a mathematician, most physicists do not hesitate to use similar reasoning. A rigorous proof appeals to an auxiliary theorem, namely the mean value theorem, sets $f(\xi) = f(x + \theta \Delta x)$, where $0 \le \theta \le 1$, and proceeds as follows:

$$K(x,\Delta x) = \int_x^{x+\Delta x} f(\xi)\, d\xi = f(x + \theta \Delta x) \int_x^{x+\Delta x} d\xi = f(x)\Delta x,$$

and the result stated in Eq. (3.17) is recovered upon taking the limit $\Delta x \to 0$;

$$\frac{d\Phi(x)}{dx} = \lim_{\Delta x \to 0} \frac{K(x,\Delta x)}{\Delta x} = \lim_{\Delta x \to 0} \frac{f(x + \theta \Delta x)\Delta x}{\Delta x} = f(x).$$

44

It is left for the reader to verify that

$$\frac{d\Psi(x)}{dx} = \frac{d}{dx}\int_x^b f(\xi)\,d\xi = -f(x).$$
(3.18)

Relation between an integral with a variable endpoint and a primitive of the integrand

Since we are still following the Leibnizian tradition, we will rephrase Eq. (3.17) as

$$d\Phi(x) = f(x)\,dx,$$
(3.19)

and integrate it to get

$$\Phi(x) = \int_a^x f(\xi)\,d\xi = F(x) + A,$$
(3.20)

and, instead of calling A an arbitrary constant, we will say that A is a constant yet to be determined. In order to determine A, we put $x = a$ in Eq. (3.20), and obtain thereby

$$\Phi(a) = \underbrace{\int_a^a f(\xi)\,d\xi}_{=0} = F(a) + A, \quad \text{or} \quad A = -F(a),$$
(3.21)

and we get, if we insert the above expression for A into Eq. (3.20),

$$\Phi(x) = \int_a^x f(\xi)\,d\xi = F(x) - F(a).$$
(3.22)

If we replace x by b in the above equation, we would recover the Newton-Leibniz formula, but it seems worthwhile to take a slightly longer route. The reasoning which led us from Eq. (3.17) to Eq. (3.22), when applied to Eq. (3.18), gives

$$\Psi(x) = \int_x^b f(\xi)\,d\xi = F(b) - F(x).$$
(3.23)

By adding the last two equations, we arrive at the Newton-Leibniz formula:

$$\int_a^b f(\xi)\,d\xi = \int_a^x f(\xi)\,d\xi + \int_x^b f(\xi)\,d\xi = F(b) - F(a).$$
(3.24)

More on definite integration in d-Calculus

The lower and upper limits for a definite integral used to be stated in words before Fourier decided to put them as sub- and superscripts [95]. The appearance of dx in $\int (\cdots) \, dx$ needs no explanation, because the integral sign *must* act on (and, therefore, must be followed by) a differential. However, one is under no obligation to place it at the end, since it always arises as a factor in a simple product. Experience shows (see below) that sometimes it is more convenient to place the differential immediately after the integral sign.

Indeed, Fourier himself did as he pleased; in fact, the first displayed equation containing an integral (which occurs at the end of §179) has the following appearance [95]:

$$y = \frac{1}{2} \int \left(dx. \frac{\sin. \, 2 \, m \, x}{\cos. \, x} \right).$$

This is how the formula appeared in the original French version of Fourier's book (1822). If you look at the footnote of the 1813 article of John Herschel [96], in which he proposes a new notation for inverse functions, you will find $\sin. x$, $\cos. x$, $\tan. x$. $\log. x, \ldots$, and also the new symbols $\sin.^{-1}$, $\cos.^{-1}$, $\log.^{-1}, \ldots$; you will also notice that Herschel used the symbol c for the constant that we (and even Fourier) represent(ed) by the symbol e. In the 1878 English translation of Fourier's book, the above integral appears in the form

$$y = \frac{1}{2} \int \left(dx \, \frac{\sin 2mx}{\cos x} \right),$$

which will not cause the raising of any eyebrows in our time.

Fourier's footprints

An example from Fourier's celebrated book is worth examining at this point. I quote from the English translation on this occasion, and begin in the middle of §216 [97, p. 181]. Since extremely few equations are numbered in old texts, I have added my own numbers, and have also replaced, for the convenience of

46

the reader, the older symbol $\underline{n}$ by $n!$:

$$\frac{1}{2}x = \sin x - \frac{1}{2}\sin 2x + \frac{1}{3}\sin 3x - \frac{1}{4}\sin 4x + \&c. \qquad (181\text{-}1)$$

In fact, multiplying each member by dx, and integrating, we have

$$C - \frac{x^2}{4} = \cos x - \frac{1}{2^2}\cos 2x + \frac{1}{3^2}\cos 3x - \frac{1}{4^2}\cos 4x + \&c.; \qquad (181\text{-}2)$$

the value of the constant C is

$$1 - \frac{1}{2^2} + \frac{1}{3^2} - \frac{1}{4^2} + \frac{1}{5^2} - \&c.;$$

a series whose sum is known to be $\frac{1}{2}\frac{\pi^2}{3!}$. Multiplying by dx the two members of the equation

$$\frac{1}{2}\frac{\pi^2}{3!} - \frac{x^2}{4} = \cos x - \frac{1}{2^2}\cos 2x + \frac{1}{3^2}\cos 3x - \frac{1}{4^2}\cos 4x + \&c.; \qquad (181\text{-}3)$$

and integrating we have

$$\frac{1}{2}\frac{\pi^2 x}{3!} - \frac{1}{2}\frac{x^3}{3!} = \sin x - \frac{1}{2^2}\sin 2x + \frac{1}{3^2}\sin 3x - \&c.; \qquad (181\text{-}4)$$

[Remark: An attentive reader could not have failed to notice that 2^2 and 3^2 on the right hand side are misprints for 2^3 and 3^3, respectively.]

The purpose of quoting the above passage is to stress that within the framework of d-Calculus, one can only integrate a differential. Fourier is therefore obliged to multiply Eqs. (181-1) and (181-3) by dx before carrying out the integration.

A symbol with two contradictory names

Earlier (pp. 42–44), we considered the integral

$$\int_a^x f(\xi)\,d\xi = \Phi(x), \qquad (3.11)$$

but no name was given to it. Should we call it a definite integral or an indefinite integral? This is the sort of question where the author should ask the reader to count up to ten before giving an answer. Picard [98, p. 8], Hardy [59, p. 287], Piskunov [94, p. 402], and Spiegel [99, p. 83], consider it to be a definite integral, whereas De Morgan [89, pp. 99–100], Courant [93, p. 110], Franklin [100] and Taylor [101] call it an indefinite integral.

The books cited above are among those I have frequently consulted, and found to be exceptionally sound and well-written. Another book in the same category uses both labels within a span of two pages. On p. 321, Eq. (7.23) is stated [102]: "In terms of the definite integral we have

$$F(x) = \int_a^x f(t)\,dt."$$ (7.23 of Ref. [102])

At the bottom of the same page, one reads: "…the most general function $G(x)$ whose derivative is equal to $f(x)$ must be of the form

$$G(x) = \int_a^x f(t)\,dt + C,$$ (7.28 of Ref. [102])

where C is a constant." The author continues thus on the next page: "The first term on the right hand side of Eqn. (7.28) is called an indefinite integral."

A primitive with a known value at a point

Consider once again the problem of finding a solution to the differential equation

$$\frac{dy}{dx} = f(x)$$ (3.25a)

that fulfils the condition

$$y(a) = A.$$ (3.25b)

When we are given such a problem, $f(x)$ is a known function, and we will assume that we are able to find one function $F(x)$ which is such that $dF(x)/dx = f(x)$; likewise $y(a) = A$ is supposed to be a known number. The usual undergraduate approach to solving such a problem begins by integrating both sides with respect to x and writing the general solution as

$$y(x) = \int f(x)\,dx = F(x) + C.$$ (3.26)

The next step is to to determine C by imposing the given initial condition, which gives

$$A = F(a) + C, \quad \text{or} \quad C = A - F(a).$$ (3.27)

Substituting the above value of C in the general solution, one gets

$$y(x) = A + F(x) - F(a) \qquad (3.28)$$

Let us transpose the first term on the right hand side and write the above result as

$$y(x)\Big|_a^x = F(x)\Big|_a^x \quad \text{or} \quad y(u)\Big|_a^x = F(u)\Big|_a^x \qquad (3.29)$$

Since $y(u) = \int_u dy/du$ and $F(u) = \int_u dF(u)/du = \int_u f(u)$, we can express the last equality as

$$\left[\int_u \frac{dy}{du}\right]_a^x = \left[\int_u f(u)\right]_a^x \qquad (3.30)$$

By virtue of the Newton-Leibniz theorem, each side of the above equation can be written as a definite integral. We are thus entitled to write

$$\int_{u=a}^x \frac{dy}{du} = \int_{u=a}^x f(u) \qquad (3.31)$$

Since Eq. (3.31) has been derived by manipulating Eq. (3.28), both must represent the same solution of the given initial value problem.

Integration between two limits provides an alternative route to finding a primitive whose value at a point is known. The recipe is summarized in Box 3.1 in the notation of d-Calculus.

Box 3.1

An initial value problem is stated below:

$$\begin{cases} \dfrac{dy}{dx} = f(x), & (3.32a) \\[2mm] y(a) = A. & (3.32b) \end{cases}$$

To find the function y that satisfies the differential equation (3.32a) and fulfils the condition stated in Eq. (3.32b), proceed as follows:

Replace the independent variable x in Eq. (3.32a) by the dummy variable u and integrate from a to x, which gives

$$y(x) - y(a) = \int_a^x f(u)\,du \qquad (3.33a)$$

Use (3.32b) to replace $y(a)$ by A, and transpose to get

$$y(x) = A + \int_a^x f(u)\,du. \qquad (3.33b)$$

Anyone who has grasped the logic behind the above two-step procedure would normally skip the first step and write down Eq. (3.33b) directly. Normally, one would simply say "Integrate Eq. (3.32a) from a to x", leaving it to the reader to understand the need for the introduction of the dummy variable.

Differentiation of (3.33b) shows that y satisfies Eq. (3.32a); also by setting $x = a$, one gets $y(a) = A$. Thus, the right hand side of (3.33b) is the desired primitive.

The fundamental theorem of calculus

The fundamental theorem of calculus is usually presented in two or more parts, not necessarily in the same order by different authors. One part deals with the relation that has been called above the Newton-Leibniz formula, (see pp. 41–42). The other part(s) of the theorem involve statement(s) concerning what we have called an *integral with a variable endpoint*, in particular Eqs. (3.11) and (3.13).

The designation *Newton-Leibniz formula* is used almost exclusively by authors affiliated with the Soviet bloc, and none of the books (known to me) where this term is employed speaks of the fundamental theorem.

Substitution in a definite integral

Let us recall Eq. (2.28), and state it in a different notation:

$$\int f(x)\,dx = \int f[\phi(u)]\,\phi'(u)\,du, \quad \text{if} \quad x = \phi(u). \qquad (3.34)$$

It follows from the above equation that

$$\left[\int f(x)\,dx\right]_{x_1}^{x_2} = \left[\int f[\phi(u)]\,\phi'(u)\,du\right]_{u_1}^{u_2}, \qquad (3.35)$$

where the symbols have the meanings already assigned to them, and it has been taken for granted that the proviso "if $x = \phi(u)$" has been met. If we apply the Newton-Leibniz formula to each side we arrive at the result stated below:

$$\int_{x_1}^{x_2} f(x)\,dx = \int_{u_1}^{u_2} f[\phi(u)]\,\phi'(u)\,du. \qquad (3.36)$$

The notation has been chosen so as to make Eq. (3.36) almost identical with the first equation (unnumbered) on p. 65 of Wilson's *Advanced Calculus* [68]; the only difference is that Wilson uses t instead of u. I will use u when quoting Wilson, who writes: "if $x = \phi(u)$, then $dx = \phi'(u)\,du$ and apparently" the left hand integral is transformed into that on the right hand side. "But this substitution is too hasty; for the dx written in the integrand is really Δx, which differs from dx by an infinitesimal of higher order when x is not the independent variable." Wilson conducts a careful examination of the conditions for the validity of the transformation formula and concludes: "Hence the change of variable suggested by the hasty substitution is justified."

Chapter 4

Two Variables: Differentiation

Here the notation is rather treacherous. On the left-hand side of (7a) we think of f as being actually expressed in terms of x and y, the variable z having been replaced by its equivalent in terms of these variables. Then we imagine that y is held constant in the x differentiation. On the right-hand side f is expressed in its original form, in terms of x, y, and z. The first term on the right *is again calculated with the other explicit variables y and z held constant,* and it represents the contribution due to the explicit variation of x. The other term adds the contribution of the only intermediate variable z. Since z depends only on x and y, the last subscript may be omitted without confusion. However, the subscript on the left is clearly essential. (Emphasis in the original.)

F. B. Hildebrand [103, p. 337]

DIFFERENTIATION of functions of several independent variables will be discussed rather briefly in this chapter, which is concerned with partial differentiation. The main purpose here is to introduce the notation and some of the basic relations that will be used later. There is no need to go into much detail, because the reader is already supposed to be familiar with the topic.

Definitions

A caution that the notation of partial differentiation can confound the beginner has already appeared in the passage cited at the head of this chapter. There is a general impression among those studying and teaching calculus that the adoption of the symbol "∂" has put an end to the mess that once existed. I would like to sum up the situation by quoting two excellent works published in the early part of the 20th century, and begin by quoting (verbatim, apart from one addition that is pointed out later) from the first edition Hardy's *Course of Pure Mathematics* [42, pp. 262–3]:

134.   Differentiation of functions of several variables.
Suppose that $f(x,y)$ is a function of two* real variables x and y, and that the limits

$$\lim_{h \to 0}\{f(x+h,y)-f(x,y)\}/h, \quad \lim_{k \to 0}\{f(x,y,+k)-f(x,y)\}/k$$

both exist; i.e. that $f(x,y)$ has a derivative with respect to x, which we have agreed to denote by df/dx or $D_x f(x,y)$, and also a derivative with respect to y, denoted similarly by df/dy or $D_y f(x,y)$. Another notation which it is natural to use is that of

$$f'_x(x,y), \quad f'_y(x,y)$$

or simply f'_x, f'_y or simply f_x, f_y. It is usual to call these differential coefficients *partial differential coefficients*, and the process of forming them *partial differentiation*. We shall not, however, adopt this method of expression, which is apt to be misleading, and to lead the reader to imagine that there is some mystery about the process of 'partial differentiation.' In point of fact it is exactly the same process as 'ordinary differentiation.' The only novelty which occurs is the presence of a second variable y in $f(x,y)$. But the variation of y has nothing to do with that of x, and there is no possible ambiguity about the meaning of $df/dx, df/dy$, which are formed in precisely the same way as $f'(x)$ or df/dx in Ch. VI.
    It is also usual to write

$$\frac{\partial f}{\partial x}, \quad \frac{\partial f}{\partial y}$$

for $df/dx, df/dy$, when f is a function of two variables: but, for the reasons indicated above, we shall not make any use of this notation.

---

*Hardy's note: The new points which arise when we consider functions of several variables are illustrated sufficiently when there are two variables only. We take for granted the obvious generalisations of our theorems to three or more variables.

In the third edition (1921) Hardy came round to using the notation $\partial f / \partial x$, etc. He must have also realized that though the section heading speaks of several variables, the discussion did not consider more than two. Instead of changing "several" to "two" in the heading, he opted for a couple of insertions, the first of which was an introductory sentence at the start of the section: "So far we have been concerned exclusively with functions of a single variable x, but there is nothing to prevent us applying the notion of differentiation to functions of several variables x, y, ..." The second insertion is the footnote already incorporated in the above quote.

The partial x-derivative of $z = f(x,y)$ can be represented by any member of the list displayed below (with $w = f$ or z):

$$\left(\frac{\partial w}{\partial x}\right)_y, \ \left(\frac{dw}{dx}\right)_y, \ (\mathscr{D}_x w)_y, \ (D_x w)_y, \ \frac{\partial w}{\partial x}, \ \mathscr{D}_x w. \qquad (4.1)$$

It will be helpful to use the modifier *active* for the variable with respect to which differentiation is carried out, and *passive* for the remaining variable(s). When there are only two independent variables, the reader is usually able to infer, on the basis of the context, the identity of the passive variable; for instance $\partial x / \partial r$ would be interpreted as the r-derivative of $x = r \cos \theta$ with θ held constant. However, occasionally there do arise situations, involving a change form one pair of variables to another, when ambiguity (and errors) can be avoided only by specifically indicating the passive variable (see below). This is usually done by enclosing the partial derivative in parentheses and affixing the symbol for the passive variable as a subscript on the closing bracket. With this addition, an unexpected passive variable will be represented, using the prevalent notation, by the symbol

$$\left(\frac{\partial x}{\partial r}\right)_y,$$

which offers no advantage over

$$\left(\frac{dx}{dr}\right)_y,$$

because both symbols require the same number of characters;

54

when there is only one passive variable, and its identity is self-evident, $\partial w/\partial x$ is to be preferred, because it uses fewer characters than $(dw/dx)_y$, and likewise for $\mathscr{D}_x w$ and $(D_x w)_y$. Let us turn to E. B. Wilson for some sound advice [68, p. 94]:

> As a matter of fact, it would probably be impossible to devise a simple notation for partial derivatives which should be satisfactory for all purposes. The only safe rule to adopt is to use a notation which is sufficiently explicit for the purposes in hand, and at all times to pay careful attention to what the derivative actually means in each case.

The following excerpt will show how Wilson handled various symbols [68, p. 123]:

> Consider the case $F(x,y,z) = 0$ and form the differential.
>
> $$dF(x,y,z) = F'_x\, dx + F'_y\, dy + F'_z\, dz = 0. \qquad (123{:}8)$$
>
> If z is to be the dependent variable, the partial derivative of z by x is found by setting $dy = 0$ so that y is constant. Thus
>
> $$\frac{\partial z}{\partial x} = \left(\frac{dz}{dx}\right)_y = -\frac{F'_x}{F'_z} \quad\text{and}\quad \frac{\partial z}{\partial y} = \left(\frac{dz}{dy}\right)_x = -\frac{F'_y}{F'_z} \qquad (123{:}9)$$
>
> are obtained by ordinary division after setting $dy = 0$ and $dx = 0$ respectively. If this division is to be legitimate, F'_z must not vanish at the point considered.

More on symbology of partial derivatives

John Perry (1850–1920) was an engineer and educator, who campaigned tirelessly for improvements in the teaching of mathematics, especially to engineers; according to one obituary [104], he "exerted a profound influence on the mathematical teaching of recent years". In 1902, he wrote a letter to *Nature* [105], the text of which is reproduced below:

Symbol for Partial Differentiation.

In my college days we used the symbol $\left(\dfrac{du}{dx}\right)_y$ or $\left(\dfrac{du}{dx}\right)$ (if there was only one other independent variable y) as the differential coefficient when y was constant. I still keep to this symbol.

Thus, if k is a certain kind of thermal capacity, $\left(\dfrac{dk}{dt}\right)_v$ or $\left(\dfrac{dk}{dt}\right)_p$ or $\left(\dfrac{dk}{dt}\right)_\phi$ are in my thermodynamic work perfectly definite. The mathematicians have introduced the convenient symbol for a partial differential coefficient $\dfrac{\partial u}{\partial x}$ and in much work there is no doubt about the meaning. But even in hydrodynamics there is trouble. In thermodynamics there is so much trouble with this symbol that I venture to ask for help.

The German translator of one of my books uses the same symbol $\dfrac{\partial k}{\partial t}$ for each of the above quite different things. Baynes in his thermodynamics does the same, and so do all other writers; it seems to me that everybody is doing this without thought. Are they writing for the average examination man who does not need to think, or for the real student? If the letter ∂ is to be retained, would it not be possible to use $\dfrac{\partial k}{\partial_v t}$ or $\dfrac{\partial k}{\partial_p t}$ or $\dfrac{\partial k}{\partial_\phi t}$ in the above three cases? I encourage my own students to use ∂, and I speak in the interest of such men. For myself it does not much matter, as I mean to continue using the symbolism of my youth.

Shortly afterwards, Thomas Muir replied to Perry's letter, and quoted some material from a paper that was in press at the time of writing. Muir's paper, which appeared in 1904 [106], traces the history of what came to be known as the Jacobian determinant. Muir also emphasized that Jacobi "puts the whole matter in a nutshell when he says that it is not enough to specify the function to be operated on and the particular independent variable with respect to which the differentiation is to be performed, but that it is equally necessary to indicate the involved quantities which are to be viewed as constants during the operation". At this point, Muir inserts a footnote in which he suggests a notation invented by him.

Just below Muir's letter appeared Perry's response [107], most of which is quoted below:

I am glad to think that a pure mathematician sees the difficulty met with by users of mathematics. I wish that men who write to me privately would publish their remarks. One correspondent says: "I think 'the mathematicians' made a rather stupid blunder when they introduced ∂ for partial differentiation. This way: nearly all differential coefficients *are* partial; even a complete one (assumed complete) may become partial by extension of the field of operation. So an old investigation of Kelvin's, for example, using d throughout, is, by 'the mathematicians,' replaced by the same using ∂ throughout, except one or two here and there! What is the use? It gives a lot of trouble, and as printers haven't always ∂'s, or proper sized ∂'s, it makes bad work. It should have been ∂ itself that was introduced for the exceptional use, thus making next to no alteration in the classical investigations." These are, indeed, my own views, but as my pupils go forward to University examinations I advise them to adopt the fashion which is likely to please the examiners.

In thermodynamics we cannot easily adopt Mr. Muir's suggestion. Take the simplest case of unit quantity of mere fluid. v, p, t, E and ϕ are such that they are all known if any two (except in certain cases) are known. Any one may be expressed as a function of any other two. My symbol $\left(\dfrac{dE}{dv}\right)_p$ is quite definite. But to adopt Mr. Muir's suggestion I must say:—Let $E = \ldots$ altogether I must use thirty of these curious symbols instead of five common letters, and, furthermore, I must keep them all in my head.

I regret to give my readers the melancholic message that the symbology of partial differentiation is still in the same bad state, which must mean that most people see no reason to sharpen up the notation. I am not a member of the 'Complacent Club'.

New notation: total and partial differentials

The total differential df of a function $f(x,y)$ is

$$df = \left(\frac{\partial f}{\partial x}\right)_y dx + \left(\frac{\partial f}{\partial y}\right)_x dy. \qquad (4.2)$$

If we put $dy = 0$ in Eq. (4.2), we obtain

$$df = \left(\frac{\partial f}{\partial x}\right)_y dx, \qquad (4.3)$$

and if we divide throughout by dx, we get

$$\frac{df}{dx} = \left(\frac{\partial f}{\partial x}\right)_y, \qquad (4.4)$$

which makes sense if (but only if) one remembers that the derivative on the left-hand side was obtained after setting $dy = 0$. Of course, every experienced reader instinctively converts the ordinary derivative on the left-hand side to the partial derivative on the right-hand side, and never displays Eq. (4.4) to the reader, but it seems desirable to rely not on instinct and restraint but on some notational refinement.

One remedy that comes to mind is to change the left-hand side by affixing the symbol for the passive variable, and writing Eq. (4.4) as

$$\left(\frac{df}{dx}\right)_y = \left(\frac{\partial f}{\partial x}\right)_y. \tag{4.5}$$

Since this makes sense and is consistent with Eq. (4.1), we might as well apply our fix at an earlier step, and rephrase Eq. (4.3) as

$$(df)_y = \left(\frac{\partial f}{\partial x}\right)_y (dx)_y, \tag{4.6}$$

which would be compatible with Eq. (4.5) only if we go on to define

$$\frac{(df)_y}{(dx)_y} = \left(\frac{df}{dx}\right)_y. \tag{4.7}$$

Of course, one could replace $(df)_y$ with $(\partial f)_y$ and $(dx)_y$ with $(\partial x)_y$ in Eqs. (4.6) and (4.7), but the choice made above would cause the least disturbance in the existing notational system.

The notation

$$\left(\frac{\partial f}{\partial x}\right)_y = \frac{df_y}{dx_y} \tag{4.8}$$

used by Thomsen [108] and by the authors of a book on thermodynamics [109] is similar but more compact, and some readers may prefer it when there is only one passive variable (and parentheses become redundant).

The notation favoured by Wilson [68, p. 95] can be summarized by considering a function of two variables $u = f(x, y)$ and expressing its total differential df as a sum of two partial differentials $d_x f + d_y f$:

$$d_x f = d_x u = \left(\frac{\partial u}{\partial x}\right)_y dx, \quad d_y f = d_y u = \left(\frac{\partial u}{\partial y}\right)_x dy. \tag{4.9}$$

This notation was used in an extremely useful book by Sherwood and Reed [110, Ch. 4], wherein the application of partial differentiation to thermodynamics is handled with great care.

Cancellation of partial differentials

On dividing Eq. (4.6) by $(df)_y$, one gets

$$1 = \left(\frac{\partial f}{\partial x}\right)_y \cdot \frac{(dx)_y}{(df)_y} = \left(\frac{\partial f}{\partial x}\right)_y \cdot \left(\frac{\partial x}{\partial f}\right)_y, \tag{4.10}$$

or, equivalently

$$\left(\frac{\partial f}{\partial x}\right)_y = \frac{1}{\left(\dfrac{\partial x}{\partial f}\right)_y}. \tag{4.11}$$

Suppose we have two derivatives such as $(\partial x/\partial u)_z$, $(\partial y/\partial u)_z$, taken with respect to the same independent variable and with the same passive variable. Since z is held fixed in both cases, the two partial derivatives act like ordinary derivatives with respect to the variable u. But for ordinary derivatives we should have

$$\frac{du/dz}{dy/dz} = \frac{du}{dy}. \tag{4.12}$$

Thus, in this case we have

$$\frac{\left(\dfrac{\partial u}{\partial z}\right)_w}{\left(\dfrac{\partial y}{\partial z}\right)_w} = \frac{\dfrac{(du)_w}{(dz)_w}}{\dfrac{(dy)_w}{(dz)_w}} = \frac{(du)_w}{(dy)_w} = \left(\frac{\partial u}{\partial y}\right)_w. \tag{4.13}$$

Change of variables: all at once

The pair of equations

$$u = \phi(x,y), \qquad v = \psi(x,y), \tag{4.14}$$

defines, in general, a transformation or mapping which establishes a correspondence between points in the xy and uv planes.

The mapping is designated as one-to-one, if to each point in the uv plane there corresponds one and only one point in the xy plane, and vice versa. In this case, we can also express (x, y) in terms of the (u, v) as

$$x = X(u, v), \qquad y = Y(u, v). \tag{4.15}$$

It will be convenient to call (x, y) the old and (u, v) the new variables. As each old variable will have a first partial derivative with respect to a new variable, we have altogether four such derivatives, all of which enter in the following 2×2 determinant:

$$\Delta_1 = \begin{vmatrix} \dfrac{\partial x}{\partial u} & \dfrac{\partial x}{\partial v} \\[2mm] \dfrac{\partial y}{\partial u} & \dfrac{\partial y}{\partial v} \end{vmatrix}. \tag{4.16}$$

Since the passive variable in a partial u-derivative (meaning a partial derivative with respect to u) is v and vice versa, it is not necessary, nor customary, to specify the passive variable, and we will follow this custom. Similar comments apply to the first partial derivatives of new variables with respect to the old variables, which make up the following determinant

$$\Delta_2 = \begin{vmatrix} \dfrac{\partial u}{\partial x} & \dfrac{\partial u}{\partial y} \\[2mm] \dfrac{\partial v}{\partial x} & \dfrac{\partial v}{\partial y} \end{vmatrix}. \tag{4.17}$$

Let us also write down the total differentials of the old and new variables:

$$dx = \frac{\partial x}{\partial u} du + \frac{\partial x}{\partial v} dv, \qquad dy = \frac{\partial y}{\partial u} du + \frac{\partial y}{\partial v} dv; \tag{4.18}$$

$$du = \frac{\partial u}{\partial x} dx + \frac{\partial u}{\partial y} dy, \qquad dv = \frac{\partial v}{\partial x} dx + \frac{\partial v}{\partial y} dy. \tag{4.19}$$

In Eqs. (4.16) and (4.18), x is to be identified with X, and y with Y; in other words,

$$\frac{\partial x}{\partial u} \text{ means } \frac{\partial X}{\partial u}, \quad \frac{\partial y}{\partial u} \text{ means } \frac{\partial Y}{\partial u}, \text{ etc.}$$

Similarly, in Eqs. (4.17) and (4.19), u is to be identified with ϕ, v with ψ.

We can solve the pair of Eqs. (4.18) to get

$$du = \frac{\mathscr{D}_v y \, dx - \mathscr{D}_v x \, dy}{\Delta_1},$$

$$dv = \frac{-\mathscr{D}_u y \, dx + \mathscr{D}_u x \, dy}{\Delta_1}, \tag{4.20}$$

and the pair of Eqs. (4.19) provide analogous relations:

$$dx = \frac{\mathscr{D}_y v \, du - \mathscr{D}_y u \, dv}{\Delta_2},$$

$$dy = \frac{-\mathscr{D}_x v \, du + \mathscr{D}_x u \, dv}{\Delta_2}. \tag{4.21}$$

Change of variables: one at a time

In the last section, we hopped directly from the xy-plane to the uv-plane. We will now make the same transition in two steps, changing one coordinate at each step. This can be done in four different ways:

$$\begin{aligned}
1: \quad & (x,y) \to (u,y) \to (u,v) & (4.22)\\
2: \quad & (x,y) \to (v,y) \to (v,u) & (4.23)\\[4pt]
3: \quad & (x,y) \to (x,u) \to (v,u) & (4.24)\\
4: \quad & (x,y) \to (x,v) \to (u,v) & (4.25)
\end{aligned}$$

It will be enough for us to look at the first option, and here too only at the first step, where we go from (x,y) to (u,y) by expressing x in terms of u and y, say as $x = \chi(u,y)$. A partial derivative of interest in this context will turn out to be

$$\left(\frac{\partial x}{\partial u}\right)_y, \tag{4.26}$$

and it is understood that the dependent variable x on in the above derivative stands for the function χ, and that the above derivative is another way of writing $(\mathscr{D}_u\chi)_y$.

Let us consider, as a concrete example, the change from (x,y) to (r,y), where r and y are considered as independent coordinates; with this choice, the remaining variables, x and θ can each be expressed in terms of r and y. Thus we have

$$x = \chi(r,y) = \sqrt{r^2 - y^2}, \quad \theta = \Theta(r,y) = \sin^{-1}\frac{y}{r}. \quad (4.27)$$

We can now differentiate χ (or Θ) with respect to r while keeping y constant (or the other way round). For example,

$$\left(\frac{\partial\chi}{\partial r}\right)_y = \left(\frac{\partial x}{\partial r}\right)_y = \frac{r}{\sqrt{r^2-y^2}} = \frac{1}{\cos\theta} = \frac{r}{x}. \quad (4.28)$$

We notice that when r is treated as a function of x and y by writing

$$r = R(x,y) = \sqrt{x^2+y^2}, \quad (4.29)$$

and R is differentiated with respect to x, keeping y constant, we get

$$\left(\frac{\partial R}{\partial x}\right)_y = \left(\frac{\partial r}{\partial x}\right)_y = \frac{x}{\sqrt{x^2+y^2}} = \frac{x}{r} = \cos\theta, \quad (4.30)$$

and we see that

$$\left(\frac{\partial r}{\partial x}\right)_y \cdot \left(\frac{\partial x}{\partial r}\right)_y = 1, \quad (4.31)$$

or, equivalently,

$$\left(\frac{\partial x}{\partial r}\right)_y = \left[\left(\frac{\partial r}{\partial x}\right)_y\right]^{-1}. \quad (4.32)$$

Some readers, because they think of (x,y), (r,θ) and (u,v) as pairs created in heaven, might say that when the active and passive variables are a heavenly pair, the identity of the passive partner need not be shown explicitly in a partial derivative. I see no reason to displease these romantics, and will continue to write (except when prevented by some other consideration)

$$\left(\frac{\partial f}{\partial x}\right)_y \quad \text{as} \quad \left(\frac{\partial f}{\partial x}\right) \quad \text{or as} \quad \mathscr{D}_x f, \quad (4.33)$$

and will replace, whenever desirable, an awkward derivative like

$$\left(\frac{\partial x}{\partial r}\right)_y \quad \text{by} \quad \left[\left(\frac{\partial r}{\partial x}\right)\right]^{-1} \quad \text{or by} \quad \frac{1}{\mathscr{D}_x r}. \qquad (4.34)$$

Functional determinants or Jacobians

In order to conform with standard practice, the determinants which have been denoted above by the symbols Δ_1 and Δ_2 will now be called *Jacobians*, and will be represented (except when brevity is desired) by the alternative symbols shown below:

$$\Delta_1 = \begin{vmatrix} \dfrac{\partial x}{\partial u} & \dfrac{\partial x}{\partial v} \\[2mm] \dfrac{\partial y}{\partial u} & \dfrac{\partial y}{\partial v} \end{vmatrix} \qquad (4.35)$$

$$= \frac{\partial(x,y)}{\partial(u,v)} \qquad (4.36)$$

$$= J\left(\frac{x,y}{u,v}\right). \qquad (4.37)$$

$$\Delta_2 = \begin{vmatrix} \dfrac{\partial u}{\partial x} & \dfrac{\partial v}{\partial x} \\[2mm] \dfrac{\partial u}{\partial y} & \dfrac{\partial v}{\partial y} \end{vmatrix} \qquad (4.38)$$

$$= \frac{\partial(u,v)}{\partial(x,y)} \qquad (4.39)$$

$$= J\left(\frac{u,v}{x,y}\right). \qquad (4.40)$$

The Jacobian has some remarkable mathematical properties, of which only three will be mentioned here. The Jacobian for the (x,y) to (u,v) transformation is the reciprocal of the Jacobian for the converse transformation:

$$J\left(\frac{u,v}{x,y}\right) \cdot J\left(\frac{x,y}{u,v}\right) = 1. \qquad (4.41)$$

Furthermore, we have the product rule,

$$J\left(\frac{u,v}{z,w}\right) \cdot J\left(\frac{z,w}{u,v}\right) = J\left(\frac{x,y}{u,v}\right), \qquad (4.42)$$

and the relation between a partial derivative and a Jacobian:

$$J\left(\frac{x,y}{u,y}\right) = \left(\frac{\partial x}{\partial u}\right)_y. \qquad (4.43)$$

One sees from Eqs. (4.41) and (4.42) that the representation of the Jacobian as a fraction $\partial(x,y)/\partial(u,v)$ gives it the appearance of a 'super' two-dimensional partial derivative.

For a straightforward proof of Eq. (4.41) based on the rules governing the multiplication of determinants and partial differentiation, see Wilson's book [68, p. 132], or some other work dealing with partial differentiation; a shorter demonstration will be presented later in this chapter [Eq. (4.49)].

For reasons that will become transparent in the next chapter, we will form four products, each with two factors. Both factors in a product will be the partial derivative of an old variable with respect to a new variable, the passive variable being the other new variable in only one derivative. Each product can also be transformed into a quotient. All four products P_i (and equivalent quotients $Q_i = P_i$) are listed below:

$$P_1 \equiv (\mathscr{D}_u x)_y \, (\mathscr{D}_v y)_u, \quad Q_1 \equiv \frac{\mathscr{D}_v y}{\mathscr{D}_x u} \qquad (4.44a)$$

$$P_2 \equiv (\mathscr{D}_u y)_x \, (\mathscr{D}_v x)_u, \quad Q_2 \equiv \frac{\mathscr{D}_v x}{\mathscr{D}_y u} \qquad (4.44b)$$

$$P_3 \equiv (\mathscr{D}_v x)_y \, (\mathscr{D}_u y)_v, \quad Q_3 \equiv \frac{\mathscr{D}_u y}{\mathscr{D}_x v} \qquad (4.44c)$$

$$P_4 \equiv (\mathscr{D}_v y)_x \, (\mathscr{D}_u x)_v, \quad Q_4 \equiv \frac{\mathscr{D}_u x}{\mathscr{D}_y v} \qquad (4.44d)$$

We will now investigate the relation between these products and $\Delta_1 = \partial(x,y)/\partial(u,v) = J(x,y/u,v)$. Set $dy = 0$ in the first of Eqs. (4.20) to get

$$(du)_y = \frac{(\partial_v y)_u \, (dx)_y}{\Delta_1}, \quad \therefore \Delta_1 = (\partial_v y)_u \underbrace{\frac{(dx)_y}{(du)_y}}_{=(\mathscr{D}_u x)_y} = P_1. \qquad (4.45)$$

On the other hand, putting $dx = 0$ in the first of Eqs. (4.20) and taking similar steps leads to the result

$$P_2 = -\Delta_1. \tag{4.46}$$

Similarly, putting $dx = 0$ or $dy = 0$ in the second of Eqs. (4.20), yields the other two products:

$$P_3 = -\Delta_1, \quad P_4 = \Delta_1. \tag{4.47}$$

An analogous treatment (that is, setting $du = 0$ or $dv = 0$), when applied to the pair of Eqs. (4.21), leads to the following results:

$$\mathcal{M}_1 \equiv (\mathcal{D}_x u)_y\, (\mathcal{D}_y v)_u = \Delta_2, \quad \mathcal{M}_2 \equiv (\mathcal{D}_y u)_x\, (\mathcal{D}_x v)_u = -\Delta_2$$
$$\mathcal{M}_3 \equiv (\mathcal{D}_x v)_y\, (\mathcal{D}_y u)_v = -\Delta_2, \quad \mathcal{M}_4 \equiv (\mathcal{D}_y v)_x\, (\mathcal{D}_x u)_v = \Delta_2 \tag{4.48}$$

A scrutiny of the composition of the product $\mathcal{M}_i$ with that of its counterpart $\mathcal{P}_i$ will show that $\mathcal{M}_i \mathcal{P}_i = 1$ for every value of i (1–4). Whence follows the relation

$$\Delta_1 \Delta_2 = 1. \tag{4.49}$$

So far, we have studiously avoided reference to one or more of the well-known properties of the Jacobian. It is interesting nonetheless to show that the relation $\mathcal{P}_i = J(x,y/u,v)$ can be derived by making use of Eqs. (4.42) and (4.43). Taking $\mathcal{P}_1$ as a concrete example, we express its two factors as Jacobians by invoking Eq. (4.43):

$$\left(\frac{\partial x}{\partial u}\right)_y = J\left(\frac{x,y}{u,y}\right), \quad \left(\frac{\partial y}{\partial v}\right)_u = J\left(\frac{u,y}{u,v}\right).$$

Multiplying the two derivatives and using Eq. (4.42), one gets

$$\mathcal{P}_1 \equiv \left(\frac{\partial x}{\partial u}\right)_y \cdot \left(\frac{\partial y}{\partial v}\right)_u = J\left(\frac{x,y}{u,y}\right) \cdot J\left(\frac{u,y}{u,v}\right) = J\left(\frac{x,y}{u,v}\right).$$

Jacobians in thermodynamics

Thermodynamics is that field of science to which one may apply Mark Twain's quip without making too many substitutions in the master author's text: "There is something fascinating about science. One gets such wholesale returns of conjecture out of such a trifling investment of fact." Starting with a handful of formulas, one can use partial differentiation to generate a myriad of thermodynamic relations, correct but not necessarily useful.

Thermodynamics is also the field where some authors frankly acknowledge that the standard notation of partial differentiation needs some extension (§§ "More on symbology ..." pp. 54–56 and "New notation ..." pp. 56–58).

Since the topic has been treated adequately by Thomsen in a pedagogical journal [108], there is no need to go into details here, especially because his nomenclature is so close to that adopted above; the reader is reminded that Thomsen writes dx_z instead of $(dx)_z$. One of his incidental remarks might be of interest to teachers of the subject: "While the treatment is not dependent on explicit use of Jacobians, it is shown in passing how they occur in solving for the desired derivatives."

Those interested in reading about the utility of Jacobians in thermodynamic calculations will find several key references in Thomsen's paper. Noteworthy among the articles that appeared after the publication of Thomsen's article is a comparatively recent compilation of useful formulas [111].

Bridgman's notation: a word of caution

On of the first attempts to systematize the presentation of thermodynamic derivatives was made (1914) by P. W. Bridgman, a leading physicist of his time and winner of the 1946 Nobel prize for his work in high pressure physics. He considered a group of 72 derivatives, and denoted by p the variable that is kept constant in all derivatives within the group, any one of which is of the form $(\partial x / \partial y)_p$, where x and y are any two of the other variables of interest. He began by introducing a new notation, which appears to be identical with that adopted above: "Now let us write, merely as a matter of notation, $\left(\dfrac{\partial x}{\partial y}\right)_p \equiv \dfrac{(\partial x)_p}{(\partial y)_p}$." Although Bridgman referred to $(\partial x)_p$ and $(\partial y)_p$ in this identity as partial differentials, they turn out to be an abbreviated expressions for derivatives with respect to an auxiliary variable [112]:

To prove the possibility of splitting up a derivative in the way above is simple. We have the mathematical identity $\left(\dfrac{\partial x}{\partial y}\right)_p \equiv \left(\dfrac{\partial x}{\partial \alpha}\right)_p \Big/ \left(\dfrac{\partial y}{\partial \alpha}\right)_p$, where α, is any variable, not even necessarily one of the fundamental ten, which remains the same throughout the group of 72. If therefore, we replace $(\partial x)_p$ by $\left(\dfrac{\partial x}{\partial \alpha}\right)_p$, and similarly $(\partial y)_p$ by $\left(\dfrac{\partial y}{\partial \alpha}\right)_p$, we shall always obtain the right answer when we take the ratio of any two such functions to find the derivative. It is especially to be noticed that $(\partial x)_p$ is not equal to $\left(\dfrac{\partial x}{\partial y}\right)_p$; in fact, $(\partial x)_p$ in general does not have the same dimensions as $\left(\dfrac{\partial x}{\partial y}\right)_p$. The finite functions replacing the differentials have meaning only when the *ratio* of two is taken. (Emphasis in the original.)

Bridgman's treatment of the problem is now mainly of historical interest, having been superseded by later work in the field. One of the reasons for discussing it here is to alert the reader to the fact that Bridgman's symbol $(\partial x)_p$ and its name *partial differential* stand for a partial derivative. Dearden has attempted to revive Bridgman's approach in a 1995 article [113]; using $\{P, \beta\}$ instead of $\{p, \alpha\}$ and retaining Bridgman's nomenclature, Dearden explained the notation:

The procedure used is to divide the derivatives into groups, each group defined by the variable held constant. Now consider a derivative such as $(\partial x / \partial y)_P$, where P is constant, and x and y represent any two of the remaining thermal variables. By postulating the existence of an auxiliary variable β for each group, β being a function of the other variables involved, one may write

$$\left(\frac{\partial x}{\partial y}\right)_P \equiv \left(\frac{\partial x}{\partial \beta}\right)_P \left(\frac{\partial \beta}{\partial y}\right)_P \equiv \frac{(\partial x)_P}{(\partial y)_P}. \tag{1}$$

Thus the partial derivative is effectively separated into numerator $(\partial x)_P$, and denominator $(\partial y)_P$, for convenience here referred to as *partial differentials*, which may be regarded as notational abbreviations for partial derivatives involving this auxiliary variable α. (Emphasis in the original.)

The importance of indicating passive variables

As has been already stated, regardless of whether one uses the symbol ∂ or d, calculations involving partial differentiation proceed more smoothly if the passive variables are explicitly indicated, as exemplified below:

$$\left(\frac{\partial f}{\partial x}\right)_{y,z} = \left(\frac{df}{dx}\right)_{y,z}.$$

It has also been pointed out above that identifications like

$$\left(\frac{\partial f}{\partial x}\right)_y \equiv \frac{(df)_y}{(dx)_y} \quad \text{and} \quad \left(\frac{\partial f}{\partial x}\right)_{y,z} \equiv \frac{(df)_{y,z}}{(dx)_{y,z}}$$

make light work of manipulations by eliminating ambiguities. Although it is highly likely that many teachers have discovered on their own the benefits of the above identities, I have not been able to find a trace of this trick predating the 1964 paper of John S. Thomsen in which the notation

$$\left(\frac{\partial f}{\partial x}\right)_y = \frac{df_y}{dx_y} \tag{4.8}$$

is exploited for manipulating thermodynamic derivatives [108]. This paper has not broken any citation records. Among the few authors who have cited this work, one could not have read it properly, for he commented [114]: "Several other approaches have been discussed which bring varying degrees of orderliness to thermodynamic derivations while obviating the need for special tables. Such methods may involve special algebraic[11] [a set of six references, the last to Thomsen] or procedural[12] prescriptions, the former group often taking indirect advantage of the Jacobian algebra in some simplified form." The symbols Δ_1 and Δ_2 were introduced in this chapter just for the sake of avoiding the misunderstanding that "some simplified form of Jacobian algebra" is being used. Thomsen, who took direct advantage of school algebra, should be given the last word in this chapter: "In the present treatment, the Jacobian symbol is a convenient shorthand notation, but it is not essential; *all the desired results may be obtained without using the Jacobian notation.*" (Emphasis added.)

Chapter 5

Two Variables: Integration

NINE of the ten questions raised in the Preamble have been dealt with in the preceding chapters. The time has come now to confront the last nagging issue: Why do we use the absolute values of Jacobians when we change variables in a multiple integral but not in a single integral? Let us refresh our memory and recall also the words of Allendoerfer [26]: "In the theory of functions of a single variable, $y = f(x)$, a reasonable case can be made for the customary definition: $dx = \Delta x$; $dy = f'(x)\,dx$. This breaks down, however, when one extends it to functions of several variables and considers double integrals of the form $\iint f(x,y)\,dxdy$. Students are rightly baffled when they attempt to convert such an integral to polar coordinates and are told that no longer is it permissible to put $dx = -r\sin\theta\,d\theta + \cos\theta\,dr$, etc. The Jacobian must be used instead, and at this point the logical structure which was built so carefully collapses entirely. If we wish to make calculus an intellectually honest subject and not a collection of convenient tricks, it is time we made a fresh start."

Time to make a fresh start. You can say *that* again. More than fifty years have elapsed since those words were written, but it is never too late to put such good words into action.

A note on notational matters

In Chapter 4, the modifiers *active* and *passive* were introduced in the context of differentiation of a function of two (or more) independent variables. A variable was called passive if it was held constant during differentiation of the function with respect to some other variable (called the active variable). This usage will now be extended, and a variable with respect to which differentiation or integration is performed will be called active, whereas the variable that is kept constant (during differentiation or integration) will be called passive.

When we come to discuss a change of variables in a double integral, avoiding ambiguities will be far more important than using a condensed notation. No excuses will therefore be necessary for displaying the symbol of the passive variable. Let us recall, therefore, the notation introduced in Chapter 4.

$$(df)_z = (\partial f)_z, \qquad \frac{(df)_z}{(dx)_z} = \left(\frac{df}{dx}\right)_z = \frac{(\partial f)_z}{(\partial x)_z} = \left(\frac{\partial f}{\partial x}\right)_z.$$

Standard recipe for transforming a double integral

We are given the integral

$$I_2 = \int_{y_1}^{y_2} \int_{x_1}^{x_2} f(x,y)\, dx dy, \qquad (5.1)$$

and we wish to introduce two new independent variables (u,v), in terms of which the old variables may defined by the following relations

$$x = X(u,v), \qquad y = Y(u,v). \qquad (5.2)$$

When one looks at standard textbooks, one finds that the recipe for implementing the required transformation of I_2 entails three different tasks: **1.** expressing $f(x,y)$ in terms of u,v; **2.** determining the new limits; **3.** finding the relation between $dx dy$ and $du dv$. This recipe leads to the result stated below:

$$\int_{y_1}^{y_2} \int_{x_1}^{x_2} f(x,y)\, dx dy = \int_{v_1}^{v_2} \int_{u_1}^{u_2} f_2(u,v) \underbrace{\left| J\left(\frac{x,y}{u,v}\right) \right| du dv}_{=dx dy}, \qquad (5.3)$$

Why are students baffled by the Jacobian?

When they are first introduced to the indefinite integral $\int f(x)\,dx$, students are told that dx is merely a bookkeeping device that helps one to keep track of the independent variable. We have already seen the following remark [61], p. 270: "We think of the combination $\int \ldots dx$ as a single symbol: we fill in the "blank" with the formula of the function whose antiderivative we seek. We may regard the differential dx as specifying the independent variable x both in the function $f(x)$ and its antiderivatives."

In d-Calculus, a change of variable (say from x to u) is such a trivial step that the requisite formula

$$\int f(x)\,dx = \int f[\phi(u)]\frac{dx}{du}\,du, \quad \text{if } x = \phi(u). \tag{5.4}$$

can be written down without thinking. The corresponding formula in H-Calculus, which cannot be derived (or proven) without using the chain rule [see Eqs. (2.19)–(2.22)], will be written as

$$\int f(x)\,dx = \int f[\phi(u)]\left(\frac{dx}{du}\right)du \tag{5.5}$$

$$= \int f[\phi(u)]\,D_u x\,du. \tag{5.6}$$

In Eq. (5.5), the parentheses have been inserted to remind the reader that the enclosed symbol is not to be treated as a fraction.

In dealing with a double integral, the so-called *area element* $dxdy$ itself becomes the object of cardinal importance, which is a sufficiently significant departure from the one-variable case to cause dismay, which soon turns into distrust when undergraduates consider the transformation from cartesian coordinates (x,y) to plane polar coordinates (r,θ), and find that

$$dx = dr\cos\theta - r\sin\theta d\theta, \qquad dy = dr\sin\theta + r\cos\theta d\theta;$$

$$dxdy = \underbrace{rdr(\cos^2\theta - \sin^2\theta)}_{\text{principal part}} + \sin\theta\cos\theta\left[(dr)^2 + (rd\theta)^2\right]$$

$$= rdr(\cos^2\theta - \sin^2\theta). \tag{5.7}$$

Most of them would willingly give their right arm to switch the minus sign on the right-hand side of the last equation to a plus sign, but this exchange is not permitted in calculus.

What a thoughtful student might expect/think/do

Let us look once more at Eq. (5.5), and agree to call (dx/du) the transformation factor for the $x \to u$ substitution in the single-variable case. A user of H-Calculus who understands the logical basis behind Eq. (5.5) and is familiar with partial differentiation might hypothesize that, when two substitutions are made in a double integral over x and y, the overall transformation factor would turn out to be a product of two factors, each factor being a partial derivative of an old variable with respect to a new variable, as shown below

$$\left(\frac{\partial x}{\partial u}\right)_\Box \cdot \left(\frac{\partial y}{\partial v}\right)_\Box, \quad \text{or} \quad \left(\frac{\partial x}{\partial v}\right)_\Box \cdot \left(\frac{\partial y}{\partial u}\right)_\Box. \tag{5.8}$$

To go further, the student must replace the boxes by symbols for passive variables, and it is likely that (s)he will use u and v. To test this proposal, the student will probably take (u, v) to be (r, θ), and would soon be able to deduce that

$$\left(\frac{\partial x}{\partial r}\right)_\theta = \cos\theta, \ \left(\frac{\partial y}{\partial \theta}\right)_r = r\cos\theta; \quad \left(\frac{\partial x}{\partial r}\right)_\theta \left(\frac{\partial y}{\partial \theta}\right)_r = r\cos^2\theta;$$

$$\left(\frac{\partial x}{\partial \theta}\right)_r = -r\sin\theta, \ \left(\frac{\partial y}{\partial r}\right)_\theta = \sin\theta; \quad \left(\frac{\partial x}{\partial \theta}\right)_r \left(\frac{\partial y}{\partial r}\right)_\theta = -r\sin^2\theta.$$

In a scenario such as this, it would not be preposterous to suppose that this student knows nothing about two-variable integration, and that whatever (s)he lacks in book learning is more than compensated by a determination to think for oneself. Well, this student will have to conclude that at least one box in each product has been wrongly labelled.

Being thoughtful, this student will argue that the integrals

$$\int_0^a \int_0^{a^2-x^2} dy\,dx = \int_0^a \int_0^{a^2-y^2} dx\,dy = A, \tag{5.9}$$

which equal the area of a quarter of a circle of radius a in the first quadrant of the xy-plane, can be transformed into an integral in the r, θ-plane

$$\int_0^a \int_0^{a^2-x^2} dy\,dx = \int_0^{\pi/2} \int_0^a U(r,\theta)\,dr\,d\theta, \tag{5.10}$$

where U is some unknown function. The student will come to the conclusion that $U(r,\theta)$ must be independent of θ and that, since $A = \frac{1}{4}\pi a^2$, one must have $U(r) = r$, which means that

$$\left(\frac{\partial x}{\partial r}\right)_\theta \left(\frac{\partial y}{\partial \theta}\right)_r \neq U(r) \neq \left(\frac{\partial x}{\partial \theta}\right)_r \left(\frac{\partial y}{\partial r}\right)_\theta \tag{5.11}$$

The student, whom we will suppose to be not only thoughtful but also dauntless, would resort to trial-and-error, and look for choices of passive variables in Eq. (5.8) that would lead to the desired result. Her/his search would show that

$$\underbrace{\left(\frac{\partial x}{\partial r}\right)_y}_{=1/\cos\theta} \cdot \underbrace{\left(\frac{\partial y}{\partial \theta}\right)_r}_{=r\cos\theta} = r = \underbrace{\left(\frac{\partial x}{\partial r}\right)_\theta}_{=\cos\theta} \cdot \underbrace{\left(\frac{\partial y}{\partial \theta}\right)_x}_{=r/\cos\theta} \tag{5.12a}$$

$$\underbrace{\left(\frac{\partial x}{\partial \theta}\right)_y}_{=-r/\sin\theta} \cdot \underbrace{\left(\frac{\partial y}{\partial r}\right)_r}_{=\sin\theta} = -r = \underbrace{\left(\frac{\partial x}{\partial \theta}\right)_r}_{=-r\sin\theta} \cdot \underbrace{\left(\frac{\partial y}{\partial r}\right)_x}_{=1/\sin\theta} \tag{5.12b}$$

Since θ decreases when x increases (and vice versa), $\partial x/\partial\theta$ turns out to be negative whatever the passive variable. With any of the two choices made in Eq. (5.12b), the limits on the θ integrals would also be reversed, and the overall result for A will not depend on the choice of the transformation factor. The standard practice is to leave the limits in the traditional order and take the absolute value of the transformation factor.

The student now notices that all four products (of two partial derivatives) in Eq. (5.12) share the following characteristic: in one factor the passive variable is one of the old variables (x or y), and the active variable in this factor (u or v) becomes the passive variable in the other factor. The student wanted to solve this mystery, but several impending deadlines for various assignments did not allow him to continue thinking about the issue, and he decided to take the easy way out by asking one of his teachers. Even a fiercely independent undergraduate like Charles Babbage did not hesitate to consult (but only twice) his college tutors, as we will find out on reading the next section, which is an excerpt from Babbage's memoir [115, p. 27].

Babbage at Cambridge

"After a few days, I went to my public tutor Hudson, to ask the explanation of one of my mathematical difficulties. He listened to my question, said it would not be asked in the Senate House [the building where the final examination was held], and was of no sort of consequence, and advised me to get up the earlier subjects of the university studies.

"After some little while I went to ask the explanation of another difficulty from one of the lecturers. He treated the question just in the same way. I made a third effort to be enlightened about what was really a doubtful question, and felt satisfied that the person I addressed knew nothing of the matter, although he took some pains to disguise his ignorance."

One substitution at a time

We return to Eq. (5.3) now. The problems of converting $f(x,y)$ to $f_2(u,v)$ and determining the new limits of integration, being essentially of algebraical in character, will not be treated here, and we will address ourselves only to the the the question of determining the transformation factor $T(u,v)$ appearing in the right-hand side of the following equation

$$\int\int f(x,y)\,dxdy = \int\int f_2(u,v)T(u,v)\,dudv. \qquad (5.13)$$

To find $T(u,v)$, we will arrange matters so that only one variable is changed at a time. The stepwise change can be effected in four different ways:

1. $(x,y) \to (u,y) \to (u,v)$

2. $(x,y) \to (v,y) \to (v,u)$

3. $(x,y) \to (x,u) \to (v,u)$

4. $(x,y) \to (x,v) \to (u,v)$

Of the four possible sequences available to us, it will be sufficient to go through the first with meticulous care; in what follows f_1 will denote what f would become when x is eliminated in f by expressing x as a function of u and x, and f_2 will denote what

f_1 would become when y is eliminated in f_1 by expressing y as a function of u and v. We will write these functional relations as

$$x = \chi(u,y), \qquad y = Y(u,v). \qquad (5.14)$$

Since we are changing one variable at a time, we can adapt the recipe spelled out in Eq. (5.5)

$$\int f(x)\,dx = \int f[\phi(u)] \left(\frac{dx}{du} \right) dx, \quad \text{if } x = \phi(u). \qquad (5.5)$$

and applying an appropriately amended version of Eq. (5.5) to the x-integral in

$$\int (dy)_x \boxed{\int f(x,y)\,(dx)_y}.$$

The appropriate amendment mentioned above amounts to changing the ordinary derivative to a partial derivative and specifying the passive variable. In other words, we will use the relation

$$(dx)_y = \left(\frac{\partial \chi}{\partial u} \right)_y (du)_y = (\mathcal{D}_u \chi)_y\,(du)_y \qquad (5.15)$$

at the first change of the variable $(x \to u)$.

At the end of Step 1, the independent variables change from (x,y) to (u,y), with a corresponding change in the integral on the right-hand side of Eq. (5.3):

$$\int (dy)_x \int f(x,y)\,(dx)_y = \int (dy)_u \int f_1(u,y)\,(\mathcal{D}_u \chi)_y (du)_y. \qquad (5.16)$$

We now change the order of integration in the right-hand side of Eq. (5.16) and write it as

$$\int (du)_y \int f_1(u,y)\,(\mathcal{D}_u \chi)_y (dy)_u. \qquad (5.17)$$

To make the $(u,y) \to (u,v)$ change, we set $f_1(u,y) = f_2(u,v)$ and use the relation

$$(dy)_u = \left(\frac{\partial Y}{\partial v} \right)_u (dv)_u, \qquad (5.18)$$

which completes the desired transformation:

$$\iint f(x,y)\,dx\,dy = \iint f_2(u,v)\,(\mathscr{D}_u\chi)_y(\mathscr{D}_v Y)_u (du)_v (dv)_u. \quad (5.19)$$

The above result will now be stated in conventional notation; we replace χ by x, Y by y, and drop the labels of the passive variable on du and dv, to get

$$\iint f(x,y)\,dx\,dy = \iint f_2(u,v)\,(\mathscr{D}_u x)_y(\mathscr{D}_v y)_u\,du\,dv \qquad (5.20)$$

$$= \iint f_2(u,v)\,\mathcal{P}_1\,du\,dv, \qquad (5.21)$$

where

$$\mathcal{P}_1 \equiv (\mathscr{D}_u x)_y \cdot (\mathscr{D}_v y)_u \qquad (5.22)$$

is the product introduced on p. 62 and shown to be equal to the determinant $\Delta_1 = J(x,y/u,v)$ [see Eq. (4.45)]. As pointed out in Chapter 4 (p. 62), each derivative in $\mathcal{P}_1$ can be identified with a Jacobian (for a single change of variables), whose product equals the familiar Jacobian; for example

$$\mathcal{P}_1 \equiv \underbrace{\left(\frac{\partial x}{\partial u}\right)_y}_{=J(x,y/u,y)} \cdot \underbrace{\left(\frac{\partial y}{\partial v}\right)_v}_{=J(u,y/u,v)} = J\left(\frac{x,y}{u,y}\right) \cdot J\left(\frac{u,y}{u,v}\right) = J\left(\frac{x,y}{u,v}\right).$$

We will call $\mathcal{P}_1$ the transformation factor for the sequence "$x \to u$ followed by $y \to v$", hereafter abbreviated as $(x2u, y2v)$. By exchanging x and y, we obtain a second sequence $(y2u, x2v)$, with its own transformation factor, namely

$$\mathcal{P}_2 = (\mathscr{D}_u y)_x \cdot (\mathscr{D}_v x)_u = -\Delta_1. \qquad (5.23)$$

The transformation factors for the remaining two sequences, $(x2v, y2u)$ and $(y2v, x2u)$, are shown below for the sake of completeness [see Eq. (4.47)]:

$$\mathcal{P}_3 = (\mathscr{D}_v x)_y \cdot (\mathscr{D}_u y)_v = -\Delta_1, \qquad (5.24)$$

$$\mathcal{P}_4 = (\mathscr{D}_v y)_x \cdot (\mathscr{D}_u x)_v = \Delta_1 \qquad (5.25)$$

Transformation in *D*-Calculus

The circuitous route taken above can be avoided by using *D*-Calculus, where isolated differentials like dx, dy, etc. do not appear, and the formula for the one-variable case,

$$\int_x f(x) = \int_u \phi(u)\frac{dx}{du} \qquad \text{if } f(x) = \phi(u) \qquad (??)$$

can be immediately adapted to transform the double integral

$$\int_y \int_x f(x,y) \qquad (5.26)$$

in four different ways, if only one substitution is made at a time. The version given below

$$\int_y \int_x f(x,y) = \int_v \int_u f_2(u,v)\left(\frac{\partial x}{\partial u}\right)_y\left(\frac{\partial y}{\partial v}\right)_u \quad [f_2(u,v) = f(x,y)]$$

$$(5.27)$$

is arrived at through the sequence $(x,y) \rightarrow (u,y) \rightarrow (u,v)$.

Living with an asymmetrical notation

In one-variable calculus, we use dx, dy, etc. both in differentiation and in integration. In handling two-variable calculus, we use ∂ only in derivatives. What did calculus books look like before Jacobi's espousal of ∂ gained general acceptance?

Isaac Todhunter's treatise on integral calculus, first published in 1857, went through changes in 1867, 1880, and 1889, but the early part of Chapter V remained unchanged, and the following passage is from this chapter; some equation have been numbered here for the reader's convenience.

56. Let $\phi(x)$ denote any function of x; then we have seen that the integral of $\phi(x)$ is a quantity u such that $du/dx = \phi(x)$. The integral may also be regarded as the limit of a certain sum (see Arts. 2…6), and hence is derived the symbol $\int \phi(x)\,dx$ by which the integral is denoted. We now proceed to extend these conceptions of an integral to cases where we have more than one independent variable.

57. Suppose we have to find the value of u which satisfies the

equation $\dfrac{d^2u}{dydx} = \phi(x,y)$ where $\phi(x,y)$ is a function of the independent variables x and y. The equation may be written

$$\frac{d}{dy}\left(\frac{du}{dx}\right) = \phi(x,y), \tag{5.28}$$

or $$\frac{dv}{dy} = \phi(x,y), \tag{5.29}$$

if $v = \dfrac{du}{dx}$. Thus v must be a function such that if we differentiate it with respect to y, considering x as constant, the result will be $\phi(x,y)$. We may therefore put

$$v = \int \phi(x,y)\,dy, \tag{5.30}$$

that is $$\frac{du}{dx} = \int \phi(x,y)\,dy. \tag{5.31}$$

Hence u must be such a function that if we differentiate it with respect to x, considering y constant, the result will be the function denoted by $\int \phi(x,y)\,dy$. Hence

$$u = \int \left\{ \int \phi(x,y)\,dy \right\} dx. \tag{5.32}$$

The method of obtaining u may be described by saying that we first integrate $\phi(x,y)$ with respect to y, and then integrate the result with respect to x. The above expression for u may be more concisely written thus,

$$\int\int \phi(x,y)\,dy\,dx, \text{ or } \int\int \phi(x,y)\,dx\,dy.$$

On this point of notation writers are not quite uniform; we shall in the present work adopt the latter form, that is, of the two symbols dx and dy we shall put dy to the right, when we consider the integration with respect to y performed before the integration with respect to x, and vice versa.

In D-Calculus, the integral in Eq. (5.32) would be represented as follows

$$\int_x \int_y f(x,y).$$

More on notational issues

Todhunter's argument flows smoothly because the notation for partial differentiation is conveniently vague, in the sense that it omits the passive variable and uses the symbol d throughout.

If *we* make an attempt to dress the above passage in contemporary garb, we would start with replacing d by ∂ and proceed along the following lines:

$$\frac{\partial}{\partial y}\left(\frac{\partial u}{\partial x}\right) = \phi(x,y), \qquad (5.28')$$

or

$$\frac{\partial v}{\partial y} = \phi(x,y), \qquad (5.29')$$

if $v = \dfrac{\partial u}{\partial x}$. Thus v must be a function such that if we differentiate it with respect to y, considering x as constant, the result will be $\phi(x,y)$.

So far so good. If we really want to be consistent we should replace Eq. (5.30) by

$$\int \partial v = v = \int \phi(x,y)\,\partial y. \qquad (5.30')$$

Some readers might say that no one would think of being so pedantic, that it is no sin to sacrifice consistency in order to retain the comfort of using familiar notation, and they would rather go along with Todhunter and write:

$$v = \int \phi(x,y)\,dy. \qquad (5.30)$$

But to such readers, I would say "Hold your horses". Let us first look at a book on differential equations [116], in which the author integrates the equation (at the bottom of p. 28)

$$\frac{\partial F(x,y)}{\partial x} = M(x,y),$$

and writes

$$F(x,y) = \int M(x,y)\,\partial x + \phi(y).$$

The notation is explained at the top of the next page (p. 29): "where $\int M(x,y)\,\partial x$ indicates a partial integration with respect to x, holding y constant, and ϕ is an arbitrary function of y only."

As it happens, this notation for integrals is just what Jacobi recommended in the paper where he proposed that the symbols ∂ and d should be used for distinguishing partial from total derivatives (and, by implication, partial from total differentials):

"If a function contains only a single variable, one may use the character d or ∂ indifferently. The same distinction may also be made in the representation of integrations, so that the expressions

$$\int f(x,y)\,dx, \quad \int f(x,y)\,\partial x$$

have different meanings; in the former, y, and hence $f(x,y)$, are regarded as the function of x, and in the latter we have to carry out the integration solely with respect to x, taking y to be constant during the integration."

For a translation of the key passages in Jacobi's paper [117], and some background, the reader should consult Appendix E.

As has already been satated above, when the users know the independent variables in the problem under consideration and there is no ambiguity about the identity of the passive variable, the distinction between d and ∂ does serve a purpose. However, we frequently encounter situations where the identity of the passive variable(s) cannot be inferred immediately, and only a more elaborate notation can eliminate ambiguities. In fact, Jacobi himself stressed that it is not sufficient to specify the function to be differentiated and the particular independent variable with respect to which the differentiation is to be performed (here called the *active variable*), and that it is just as necessary to indicate the involved quantities which are to be viewed as constants during the differentiation (here called the *passive variables*).

In the notation used here, the partial x-derivative of a function $f(x,y,z)$ of three independent variables can be denoted by any of the following symbols,

$$\left(\mathscr{D}_x f\right)_{y,z} = \left(\frac{\partial f}{\partial x}\right)_{y,z} = \left(\frac{df}{dx}\right)_{y,z} = \left(D_x f\right)_{y,z'}$$

and the parentheses (with their subscripts) should be dropped only when there is no danger of confusing the reader. Alternatively, one may write $D_x f(x,y,z)$, etc.

Chapter 6

Iterations and Expansions

'What kind of professor are you?'

'I teach mathematics.'

'Mathematics?' said the Ataman. 'All right! Then give me an estimate of the error one makes by cutting off Maclaurin's series at the n-th term. Do this, and you will go free. Fail, and you will be shot!'

Tamm could not believe his ears, since this problem belongs to a rather special branch of higher mathematics. With a shaking hand, and under the muzzle of the gun, he managed to work out the solution and handed it to the Ataman.

'Correct!' said the Ataman. 'Now I see that you really are a professor. Go home!

George Gamow [118, p. 19]

N. B. *The anecdote, clearly spiced up by Gamow, is probably genuine; a more sober version was narrated by L. I. Vernsky, grandson of Igor Tamm* [119].

I have often used the above story as a bait for capturing the interest of my students in Taylor expansion of functions, which is the main topic of this chapter. After the Igor Tamm story had generated some mirth and created a convivial atmosphere, I used to urge my students to come up with their own answers to the questions raised at the beginning of the next section.

Series expansions

Suppose that we know the value of a function $f(x)$ at the point x_0, that we have no other information, and that we would like to estimate the value of the function at a neighbouring point $x_0 + a$. The only rational conclusion is to set $[0]f(x_0 + a) = f(x_0)$, where the left superscript has been added to indicate that this is a zero-order guess. If we happen to know not only $f(x_0)$ but also $f'(x_0)$, what would be $[1]f(x_0 + a)$, the next (first-order) estimate, and how would this first-order estimate change if we are given successively $f''(x_0)$, $f'''(x_0)$, $f^{iv}(x_0)$, etc? Finally, what might have been going on in the mind of Brook Taylor when he formulated a primitive version of the theorem that was subsequently refined and named after him?

It is easy enough to give my answers to the above questions, but the mathematical expressions will look cluttered with super and subscripts if we do not adopt a more congenial notation. I will replace x_0 by a, move the left superscript to the right and use a numerical subscript to denote the order of differentiation by introducing the symbol

$$f_n(x) \equiv D_x^n f(x) \equiv \frac{d^n}{dx^n} f(x), \qquad \left[f_0(x) \equiv f(x) \right]$$

Integration of the equation

$$\frac{df_{n-1}}{dx} = f_n(x) \tag{6.1}$$

from a to x gives

$$f_{n-1}(x) = f_{n-1}(a) + \int_a^x f_n(u)\, du. \tag{6.2}$$

For $n = 1$, this gives

$$f(x) = f(a) + \int_a^x f_1(u)\, du. \tag{6.3}$$

If the only derivative value we know is $f_1(a)$, we replace $f_1(x)$

by $f_1(a)$ and get thereby first-order approximation:

$$[1]f(x) = f(a) + f_1(a)(x - a) \qquad (6.4)$$

If $f_2(a)$ is also known, we set $n = 2$ in Eq. (6.2)

$$f(x) = f(a) + \int_a^x \underbrace{\left[f_1(a) + \int_a^u f_2(v)dv \right]}_{=f_1(u)} du \qquad (6.5)$$

$$= f(a) + (x - a)f_1(a) + \int_a^x du \int_a^u f_2(v)\, dv \qquad (6.6)$$

If $f_2(a)$ is the highest derivative available, all we can do is to set $f_2(x) = f_2(a)$, and deduce the second-order estimate:

$$[2]f(x) = f(a) + (x - a)f_1(a) + \tfrac{1}{2}(x - a)^2 f_2(a) \qquad (6.7)$$

The first two estimates, namely $[1]f(x)$ and $[2]f(x)$ are used extensively, and for practical applications, one seldom needs to go beyond the second-order approximation. If we choose to stop iteration after n steps, we get

$$f(x) = f(a) + (x - a)f_1(a) + \frac{(x - a)^2}{2}f_2(a) +$$
$$\frac{(x - a)^3}{2}f_3(a) + \cdots + \frac{(x - a)^{n-1}}{(n - 1)!}f_{n-1}(a) + R_n(x) \qquad (6.8)$$

where $R_n = \int_a^x dx_1 \int_a^x dx_2 \int_a^x dx_2 \cdots \int_a^x dx_n f_n(x_n)$

is the remainder after n terms. This is Taylor series in a finite form. If we put $a = 0$ in Eq. (6.8), we arrive at the version

$$f(x) = f(0) + xf_1(0) + \frac{x^2}{2}f_2(0) + \frac{x^3}{3!}f_3(0) + \cdots + R_n(x) \quad (6.9)$$

named after Maclaurin.

A variant of Eq. (6.8), obtained by replacing $x - a$ by h and a by x, is stated below (and will shortly be derived ab ovo):

$$f(x+h) = f(x) + hf_1(x) + \frac{h^2}{2}f_2(x) +$$
$$\frac{h^3}{3!}f_3(x) + \cdots + \frac{h^{n-1}}{(n-1)!}f_{n-1}(x) + R_n(x) \qquad (6.10)$$

The purpose of the next five sections (pp. 84–87) is to make the reader familiar with some tricks that will be needed for the derivation of Taylor's formula.

Iteration as a means of solving a simple differential equation

We will suppose that the functions log and e are known to us, and we wish to tackle the problem of finding y, given that it satisfies the differential equation (6.11a) and the subsidiary condition stated in Eq. (6.11b):

$$\frac{dy}{dx} = -ky, \qquad (6.11a)$$

$$y(0) = y_0 \qquad (6.11b)$$

We will integrate Eq. (6.11a) from 0 to x to get

$$y(x) - y_0 = -k \int_0^x dx_1\, y(x_1), \qquad (6.12a)$$

$$\therefore \quad y(x) = y_0 - k \int_0^x dx_1\, y(x_1). \qquad (6.12b)$$

Unfortunately, the integral on the right hand side involves not a *known* function but the *unknown* y. However, the foregoing manipulations have not been in vain, for they have led us to the restatement of our problem in terms of a single *integral equation*, which already incorporates the given initial condition.

Method of successive approximations
There are two ways of dealing with Eq. (6.12b). We will first use an iterative method; this means that we will make an initial guess, $[0]y(x)$, and calculate $[1]y(x)$, the first approximation to $y(x)$, by substituting $[0]y(x) = y(0)$ on the right hand side of Eq. (6.12b), and repeat this procedure indefinitely, using the

output of the n-th step as the new guess for calculating the next approximation, as shown below.

$$^{[n+1]}y(x) = y_0 - k \int_0^x dx_1 \, ^{[n]}y(x_1).$$ (6.13)

The only remaining hurdle, that of making a suitable initial guess, can be overcome simply by taking $^{[0]}y(x) = y(0) = y_0$. We will refer to the iterative method outlined in this section as the *method of successive approximations*. Since we are looking for an exact solution y, we must continue indefinitely, and see whether $^{[\infty]}y$ satisfies Eq. (6.12b). The procedure is implemented below:

$$^{[1]}y(x) = y_0 - k \int_0^x dx_1 y_0 = y_0(1 - kx),$$ (6.14)

$$^{[2]}y(x) = y_0 - k \int_0^x dx_1 y^{(1)}(x_1) = y_0(1 - kx + k^2x^2/2),$$ (6.15)

$$. . . . . .$$

$$^{[\infty]}y(x) = y_0 \left[1 - kx + \frac{(kx)^2}{2!} - \frac{(kx)^3}{3!} + \frac{(kx)^4}{4!} - \cdots \right]$$ (6.16)

$$= y_0 \sum_0^\infty \frac{(-k)^n x^n}{n!} = y_0 S_\infty(x), \text{ say.}$$ (6.17)

It is easy to verify that

$$\frac{dS_\infty}{dx} = -kS_\infty$$ (6.18)

from which it follows that $y_0 S_\infty$ is the solution to the initial value problem in Eq. (6.12a). We now drop the pretence made at the start, we will recognize that the series on the right hand side is an expansion of the exponential function e^{-kx}, and will be led to the conclusion that

$$y = y_0 e^{-kx}$$

is a solution of Eq. (6.12a), and that it satisfies the initial condition $y = y_0$ at $x = 0$, stated in Eq. (6.12b). If the above iteration had really been carried out before the discovery of the exponential function, it would have been necessary to establish that the series is convergent.

Formal iteration

Let us rephrase Eq. (6.12b) as

$$y = y_0 - kQy, \qquad (6.19)$$

with the understanding that the operator $Q \equiv \int_0^x dx_1$, gives a result only when it acts on an operand, as shown below:

$$Qy \equiv \int_0^x y(x_1)\, dx_1 \qquad (6.20)$$

The special case where the operator Q acts on a constant (say A) is worth examining:

$$QA \equiv \int_a^x A\, du = A \int_a^x du = AQ = A(x - a). \qquad (6.21)$$

Continuing in the same vein, we see that

$$Q^2 A = QQA = QA(x - a) = A\frac{(x - a)^2}{2}. \qquad (6.22)$$

With the help of the operator Q, Eq. (6.19) can be manipulated as follows:

$$y = y_0 - kQy = y_0 - kQ(y_0 - kQy) \qquad (6.23)$$
$$= (1 - kQ)y_0 + k^2 Q^2 y \qquad (6.24)$$
$$= (1 - kQ + k^2 Q^2 - k^3 Q^3 + \cdots)y_0 \qquad (6.25)$$

On the other hand, we can rearrange Eq. (6.19) as

$$(1 + kQ)y = y_0 \qquad (6.26)$$

and use this result to express y as

$$y = \frac{1}{(1 + kQ)} y_0 \equiv (1 + kQ)^{-1} y_0, \qquad (6.27)$$

but we will now have to find the meaning of the operator

$$\frac{1}{(1 + kQ)} \equiv (1 + kQ)^{-1}. \qquad (6.28)$$

Let us recall the following result

$$\frac{1}{1+x} \equiv (1+x)^{-1} = 1 - x + x^2 - x^3 + \cdots \qquad (6.29)$$

When we examine Eqs. (6.29), (6.25), and (6.19) it becomes obvious that we have no choice but to identify the operator $(1 + kQ)^{-1}$ by its binomial expansion and write

$$\frac{1}{(1+kQ)} \equiv (1+kQ)^{-1} = 1 - kQ + k^2Q^2 + \cdots \qquad (6.30)$$

An interlude

Some authors [120, 121] prefer to use x instead of the dummy variables and write

$$\int_a^x f_n(x)\,dx = f_{n-1}(x)\Big|_a^x = f_{n-1}(x) - f_{n-1}(a) \qquad (6.31)$$

$$\int_a^x \left(\int_a^x f_n(x)\,dx \right) dx = \int_a^x [f_{n-1}(x) - f_{n-1}(a)]\,dx$$
$$= f_{n-2}(x) - f_{n-2}(a) - (x-a)f_{n-1}(a) \qquad (6.32)$$

$$\int_a^x \int_a^x \int_a^x f_n(x)(dx)^3 = \left[\int_a^x \right]^3 f_n(x)(dx)^3 \qquad (6.33)$$

$$\underbrace{\int_a^x \cdots \int_a^x}_{m \text{ integrals}} f_n(x)(dx)^m = \left[\int_a^x \right]^m f_n(x)(dx)^m \qquad (6.34)$$

A painless derivation of Taylor's formula

Let us consider a function

$$u = f(x+h),$$

where f is an arbitrary differentiable function. If the argument of f is denoted by $s = x + y$, then

$$f(x+h) = f(s)$$

and the chain rule gives

$$\left(\frac{\partial u}{\partial x}\right)_h = \frac{df}{ds} \cdot \left(\frac{\partial f}{\partial x}\right)_h = \frac{df}{ds} \cdot 1 = \frac{df}{ds} \tag{6.35}$$

$$\left(\frac{\partial u}{\partial h}\right)_x = \frac{df}{ds} \cdot \left(\frac{\partial f}{\partial h}\right)_x = \frac{df}{ds} \cdot 1 = \frac{df}{ds} \tag{6.36}$$

Therefore for any choice of the function $f(x+h)$, we have

$$\left(\frac{\partial u}{\partial h}\right)_x = \left(\frac{\partial u}{\partial x}\right)_h \tag{6.37}$$

For our purpose it is more convenient to state the equation so that it resembles Eq. (6.11a):

$$\frac{\partial f(x+h)}{\partial h} = D_x f(x+h). \tag{6.38}$$

The solution is

$$f(x+h) = e^{hD_x} \phi(x) \tag{6.39}$$

where ϕ is any function of x. This is easily verified by writing the $e^{hD_x} \phi(x)$ in extenso as

$$e^{hD_x} \phi(x) = \phi(x) + hD_x\phi(x) + \frac{h^2}{2}D_x^2\phi(x) + \frac{h^3}{3!}D_x^3\phi(x) + \cdots$$

and verifying that $\mathscr{D}_h e^{hD_x} \phi(x) = D_x e^{hD_x} \phi(x)$. By putting $h = 0$ in Eq. (6.39), we see that $\phi(x) = f(x)$.

An exquisitely compact rendering of Taylor's theorem, which follows from the foregoing analysis, is stated below:

$$\boxed{f(x+h) = e^{hD_x} f(x)} \tag{6.40}$$

Though it is hardly necessary, we will now take a more leisurely route to Taylor's formula and integrate Eq. (6.38) over h. This means that $\mathscr{D}_x$ can be treated like a constant, and the equation can be considered as an analog of $dy/dk = ky$.

Iterative route to Taylor's formula

We introduce the operator

$$T(\ldots) = \int_0^h (\ldots)\, dh, \tag{6.41}$$

and apply it to the left-hand side over h from 0 to h, which yields

$$\int_0^h \frac{\partial}{\partial h} f(x+h)\, dh = f(x+h) - f(x). \tag{6.42}$$

Next, we apply T to the right hand side, pull the operator $\mathscr{D}_x$ out of the integral sign, and obtain the result stated below:

$$\int_0^h \mathscr{D}_x f(x+h)\, dh = \mathscr{D}_x T f(x+h) \tag{6.43}$$

Upon equating the right hand sides of Eqs. (6.42) and (6.43), and collecting terms containing $f(x+h)$ on one side, we get

$$[1 - \mathscr{D}_x T] f(x+h) = f(x), \tag{6.44}$$

$$\therefore f(x+h) = (1 - \mathscr{D}_x T)^{-1} f(x) \tag{6.45}$$

$$= (1 + \mathscr{D}_x T + \mathscr{D}_x^2 T^2 + \cdots) f(x) \tag{6.46}$$

$$= \left[1 + h\mathscr{D}_x + \frac{h^2}{2!} \mathscr{D}_x^2 + \cdots \right] f(x) \tag{6.47}$$

The last step follows because

$$Tf(x) = \int_0^h f(x)\, dh = f(x) \int_0^h dh = f(x)h \tag{6.48}$$

$$T^2 f(x) = Tf(x)h = f(x)Th = \int_0^h h\, dh = \frac{h^2}{2} \tag{6.49}$$

$$\cdots \quad \cdots \quad \cdots \tag{6.50}$$

$$T^n f(x) = \left[\int_0^h \right]^n f(x)(dx)^n = \frac{h^n}{n!} f(x) \tag{6.51}$$

Heaviside's view of Taylor's theorem

Heaviside made extensive use of Taylor's theorem in a most original manner. §276 of his *Electromagnetic Theory* [122] is entitled
"Taylor's Theorem in its Essentials Operationally Considered".
Some excerpts reproduced below (after some minor changes, e.g.
replacement of d by ∂, d/dt by $\mathscr{D}_t$, and of d/dx by $\mathscr{D}_x$):

> Imagine any wave form to be transmitted at a constant speed
> undeformed, as, for instance, a hump running along a flexible cord.
> Let it be as in the figure [Fig. 6], namely, a triangular hump whose
> position at the moment of time $t = 0$ is A, and at some later time t is
> B. It is to be imagined to travel at uniform speed v from left to right,
> so that the distance AB is vt, increasing uniformly with the time. If
> the base line is the axis of x, the travelling form is our "function of
> x" its value being the ordinate of the waveform.
>
> Let $f(x)$ represent the function when at A, and let it become
> $F(x)$ at . They only differ in the change of origin. Their relation to
> one another is
>
> $$F(x) = f(x - vt); \tag{77}$$
>
> that is, the value of F at x at the moment t is, by definition, the
> same as the value of f at a distance vt to the left. Equation (77)
> expresses the characteristic property of an undistorted and unat-
> tenuated wave when the speed is constant. It is a positive wave.
> When it goes the other way, it is a negative wave, and the minus
> sign must be changed to plus.
>
> Differentiate equation (77) with respect to x and t separately
> and compare them. We see that

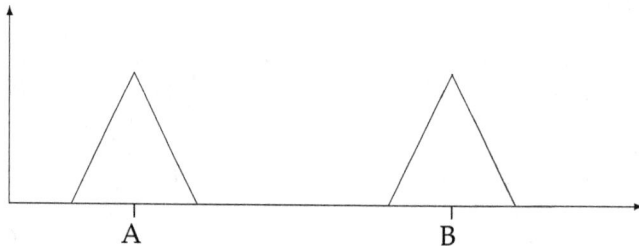

Figure 6.1: Two snapshots of a triangular waveform travelling to
the right

$$\frac{\partial F}{\partial x} = -\frac{1}{v}\frac{\partial F}{\partial t} \tag{78}$$

This is the characteristic partial differential equation of a solitary undistorted wave. It is easily understood by watching the transit of a particular part of the wave past a fixed point, and considering how the speed v of transit combined with the slope determines the time rate of increase of F at the fixed spot. Equations (78) and (77) have practically the same meaning. Thus, given (78), its solution is (77), understanding that f expresses an arbitrary function of x.

Now write (78) in the form

$$\frac{\partial F}{\partial t} = -v\mathcal{D}_x F, \tag{79}$$

where $\mathcal{D}_x$ stands for $\partial/\partial x$, and solve it as an ordinary linear equation. This makes

$$F = e^{-vt\mathcal{D}_x}\phi \tag{80}$$

where ϕ is a constant with respect to t. In our present case it is any function of x. That (80) is the solution is verifiable by its satisfying (79). What ϕ is may be seen by putting $t = 0$. Then $F = \phi$. So ϕ is the initial state of F, and is therefore the same as $f(x)$. Comparing (80) with (77), we see that the operator $e^{-vt\mathcal{D}_x}$ is the translation operator, so that

$$e^{-vt\mathcal{D}_x}f(x) = f(x - vt) \tag{81}$$

This is the universal property of the operator $e^{h\mathcal{D}_x}$ when h is a constant. What it does, when applied to a function of x, is to translate it bodily through the distance h to the left.

At this point Heaviside digresses into making some comments on the link between Fourier series and Taylor's theorem; these remarks will not make much sense to a reader who is not familiar with Heaviside's idiosyncratic treatment of differential equations arising in electrical circuit theory. I hop over this part of the text and join Heaviside where he returns to Taylor's theorem.

Equation (81), or, which is the same,

$$e^{h\mathcal{D}_x}f(x) = f(x + h) \tag{84}$$

is the well-known so-called "symbolical" form of Taylor's theorem. For it is merely the condensation of the form obtained by expanding the exponential, viz.:

$$\left(1 + h\mathscr{D}_x + \frac{h^2}{2}\mathscr{D}_x^2 + \cdots\right)f(x) = f(x+h) \tag{85}$$

The form (84) is much more convenient. It is the form that always occurs naturally in operational mathematics. That is, we get solutions of the form $e^{h\mathscr{D}_x}$ acting upon something or other, and the meaning is simply a bodily translation performed upon it.

The way followed above is not the usual way of leading to Taylor's theorem, for which see works on the Calculus. It is a quasi-physical way, by which its truth becomes obvious in a general sense. It only involves the idea of translation of a form so far as the essential part goes. But now comes the question of failures. In fact, the proofs of Taylor's theorem are largely devoted to the discussion of modified forms with remainders occurring when there is some discontinuity, involving an infinite differential coefficient. When mathematicians come to an infinity they are nonplussed, and hedge round it. They would, for example, stick at the three sharp corners in the function above used, which involve discontinuity in the slope, or infinite curvature.

But there is no difficulty of the kind in the physical way of looking at the matter, so far as the act of bodily translation is concerned. One shape of the function is just as easily conceived as another, and we are not limited to the angelically perfect function which is finite and continuous itself, and has all its derivatives finite and continuous. It was long ago remarked by Sir G. Stokes that it is important for physical mathematicians to conceive of functions wholly apart from their possible symbolical expression. The remark is particularly important in the discussion of waves and their propagation, when discontinuous functions have frequently to be dealt with.

Now (77) is unexceptionable, and requires no special interpretation involving infinities. If $f(x)$ makes a jump suddenly at any place, so does $F(x)$ precisely at a corresponding place. (The translation of the form should be thought of, not merely that of a value. If the curve is upright, the upright piece behaves just like the rest.) But (78), which is taken to be equivalent to (77), inasmuch as it utilises differential coefficients, may require interpretation. We should not say that the whole thing breaks down if the slope is infinite, because (78) becomes meaningless. It requires interpretation, but is not uninterpretable. The variations in dF/dx and those in dF/dt keep pace together precisely, and this tie does not break when they are momentarily infinite.

Although, of course, in a popular sense, infinities are immeasurable, they are not necessarily unmeasurable. A suitable standard is required to measure an infinity. Consider, for example, the case of electrification. Starting with a finite volume density, if we imagine it condensed on a surface, we have infinite volume density. But that causes no trouble. It is now to be measured by its surface density. Again, imagine it condensed from finite surface electrification to linear electrification. This means infinite surface density, and therefore doubly infinite volume density. But we now have a finite linear density, so there is no difficulty in measurement, or at any rate in the conception of the suitable way of measuring. Again, the notion of impulses in dynamics is exceedingly useful, though it involves the idea of an infinite force. We must not be afraid of infinity. If suitably measured, it may be no bigger than zero, or else quite small.

Now I think that similar ideas may with advantage be introduced into pure mathematics, to enlarge the scope of the mathematicians' investigations, and enable them to tackle infinities, instead of evading them. However, we are not concerned with the real or supposed failures in Taylor's theorem,, perhaps rather to be considered failures in the mathematical machinery. In the operational mathematics $e^{h\mathscr{D}_x}$ is the translational operator, no matter how many discontinuities may be in the way. We do not have to consider the infinite values, but just jump over.

In physical applications, say to waves, the subject of operation is real and single valued, though it need not be finite and continuous. But I do not see any reason why the translational idea should not be carried further. Let it be applied to any definite series of values, or of multiple values, or perhaps to the case of a multiple pole, involving a collection of infinities.

Anyone who has been misled to conclude that Eq. (84) was the brain child of Oliver Heaviside should note that Heaviside was almost totally immune to what is called 'precursoritis", and he seldom wasted any ink or paper or time (his own or that of his readers) in citing those who had deduced a particular result before him.

Chapter 7

Differential Equations. Part I

O NE of the aims of this book is to talk about antidifferentiation by using the vocabulary of differential equations. It will be enough for the present to state that the term *differential equation* (to be abbreviated as DE) may be applied to any equation that contains one or more derivatives. The task of finding the general antiderivative of a given function $f(x)$ can also be described as the task of finding an expression for the general solution to the following DE

$$\frac{dy}{dx} = f(x). \qquad (7.1)$$

We will not be much interested in finding the general solution to a given DE; rather, our interest lies in solving what is called an *initial value problem* (IVP), which amounts to finding a function that satisfies not only Eq. (7.1), or a closely related DE, but also an *initial condition*, which may be stated in the form $y = y_0$ at $x = x_0$.

Preliminaries

There can be no doubt that the simplest differential equation is

$$\frac{dy}{dx} = 0, \tag{7.2}$$

and that its general solution is

$$y = a, \tag{7.3}$$

where a is an adjustable constant. A particular solution can be obtained by assigning a particular value to a; we may choose, for example, $a = \pi$, or $\sqrt{2}$, or 1984, ...

The general solution to the equation

$$\frac{d^2y}{dx^2} = 0, \tag{7.4}$$

obtained by two integrations, is clearly

$$y = ax + b, \tag{7.5}$$

and it contains two adjustable constants, one for each integration step.

Integration of the equation

$$\frac{dy}{dx} = 2x \tag{7.6}$$

shows that its general solution is

$$y = \int 2x \, dx = x^2 + C, \tag{7.7}$$

where C is an undetermined constant. We call y the general solution because the value of C has not been specified.

Consider now the following IVP: Find the equation to the curve that satisfies the equation $dy/dx = 2x$ and passes through the point (0,4). The extra condition is another way of saying that $y = 4$, when $x = 0$; on inserting these values in Eq. (7.7), one gets $4 = C$, which means that the answer to our problem is $y = x^2 + 4$, which is a specific or particular solution that satisfies the given DE and the additional condition (or the initial condition).

The simplest inhomogeneous ODE

In the rest of this chapter, we will be interested in finding the general solution to an equation that is either already in the form

$$\frac{dy}{dx} = \phi(x), \tag{7.8}$$

or can be brought into the above form by very little preparatory work.

Let
$$y^\dagger = \int^\dagger \phi\, dx. \tag{7.9}$$

Then $y = y^\dagger$ is a particular solution of Eq. (7.8) because

$$\frac{dy^\dagger}{dx} = \phi. \tag{7.10}$$

Now put $y = y^\dagger + y_h$ in Eq. (7.8) and subtract Eq. (7.10) from the resulting equation; this gives

$$\frac{dy_h}{dx} = 0, \tag{7.11}$$

a homogeneous differential equation (see below) whose general solution is

$$y_h = C. \tag{7.12}$$

Hence
$$y = y^\dagger + C \tag{7.13}$$

is a solution of Eq. (7.8), and it must also be the most general solution because it contains one undetermined constant.

An equation like Eq. (7.8) is called an ordinary differential equation (ODE) because it has no partial derivatives. It is also an inhomogeneous differential equation, which is a name we give to equations whose left hand side contains the (unknown) dependent variables and its derivatives (with or without a constant term), and the right hand side contains a *known* function of the independent variable. This means that Eq. (7.8) is an inhomogeneous ordinary differential equation (or IODE, for short). Each

IODE can be turned into its homogeneous counterpart by setting the right hand side to zero; the role played by a homogeneous ordinary differential equation (HODE) will be explained presently.

It will be convenient to call Eq. (7.8) the simplest IODE, and its homogeneous counterpart, $dy/dx = 0$, the simplest HODE.

To put the contents of Eq. (7.13) into words: the complete (or general) solution to the simplest IODE, is the sum of two terms, $y^\dagger$ and C; the former, a particular solution, is any solution to the simplest IODE, whereas the latter, called the constant of integration, is the general solution to the HODE associated with the simplest IODE. It is evident that the particular solution cannot contain an arbitrary constant and it must contain the inhomogeneous term $f(x)$.

Let us recall at this point two terms which made their way, about two centuries ago, from continental Europe into English works on ODE's [123, p. 211–2], and became so well established that no attempt to replace them by more appropriate alternatives will get a fair hearing. Any solution to a given IODE is called a *particular integral* (PI), whereas the general solution of the corresponding HODE is called the *complementary function* (CF). Since we cannot but acquiesce to the existing nomenclature, we must recognize now that what was called (in the context of antidifferentiation) the particular antiderivative is, in the parlance of ODE's, a particular integral of the simplest IODE, and the constant of integration is, for the same reason, to be viewed as the general solution of the HODE associated with the simplest IODE, and called the complementary function.

The next simplest inhomogeneous ODE

We consider next the IODE

$$\frac{dy}{dx} + ky = f(x), \tag{7.14}$$

and seek both its general solution and a specific solution that satisfies the initial condition

$$y = y_0 \quad \text{when} \quad x = 0. \tag{7.15}$$

In order to reduce Eq. (7.14) to Eq. (7.8), we must somehow

eliminate the term ky on the left hand side of the former equation, which represents attenuation or damping (depending on the context). The effect of the term is seen most easily by considering the equation

$$\frac{dy}{dx} + ky = 0, \qquad (7.16)$$

which is the HODE corresponding to Eq. (7.14). The general solution of Eq. (7.16) will be shown to be

$$y = Ae^{-kx}, \qquad (A = \text{constant}). \qquad (7.17)$$

If k were equal to zero, y would remain unchanged as x is varied. Though we cannot tinker with the value of k, it is clear that the product $e^{kx}y$ would remain constant, whatever be the value of k. In the light of this observation, we should try the substitution

$$u = e^{kx}y, \qquad (7.18)$$

and calculate du/dx, which comes out to be

$$\frac{du}{dx} = ke^{kx}y + e^{kx}\frac{dy}{dx}$$
$$= e^{kx}\left(\frac{dy}{dx} + ky\right). \qquad (7.19)$$

It follows from the above that

$$\frac{dy}{dx} + ky = e^{-kx}\frac{du}{dx}. \qquad (7.20)$$

We return to Eq. (7.14), replace its left hand side by the right hand side of Eq. (7.20), and multiply by e^{kx} to remove the unwanted exponential factor on the left hand side; this leads us to

$$\frac{du}{dx} = e^{kx}f(x). \qquad (7.21)$$

We have succeeded in removing the damping term, and arrived at an equation which is of the same form as Eq. (7.8), with y replaced by $u = e^{kx}y$ and $\phi(x)$ replaced by $e^{kx}f(x)$. It follows from the definition of u and the initial condition to be imposed on y that

$$u = y_0 \quad \text{when} \quad x = 0. \qquad (7.22)$$

Though we can proceed to integrate Eq. (7.21), a beginner

might find it easier to grasp the argument if we permit ourselves a slight diversion for examining a concrete case in which $f(x)$ is chosen so as make $\phi(x) = e^{kx}f(x)$ easily integrable. The reader is advised to take $f(x) = x^2$, and go on to find the answer to the following question:

Given the general integral

$$\int x^2 e^{kx}\, dx = \frac{e^{kx}}{k^3}(k^2x^2 - 2kx + 2) + C, \qquad (7.23)$$

adjust the value of the constant C to find the particular integral which equals $6/k^3$ when $x = 0$.

Finding a solution that satisfies a given initial condition

We multiply both sides of Eq. (7.21) by dx and integrate the resulting equation to obtain the general solution to Eq. (7.21):

$$u = \int e^{kx} f(x)\, dx \qquad (7.24a)$$

$$= \underbrace{C}_{CF} + \underbrace{\int^{\dagger} e^{kx} f(x)\, dx}_{PI\ of\ Eq.\ (7.21)}. \qquad (7.24b)$$

In the context of antidifferentiation, C is called the constant of integration, but in the present context, it stands for the complementary function. Since we are interested in finding y, which equals $e^{-kx}u$, we multiply Eq. (7.24b) throughout by e^{-kx} to get the required general solution, which can be stated as:

$$y = \underbrace{C e^{-kx}}_{CF} + \underbrace{e^{-kx}\int^{\dagger} e^{kx} f(x)\, dx}_{PI\ of\ Eq.\ (7.14)}. \qquad (7.25)$$

As already stated, general solutions of ODE's will be of much less interest to us than specific solutions that satisfy additional conditions pertaining to the situations which the equations are meant to describe. The constraint for the present problem having already been specified above [see Eq. (7.15)], we now ask "What is the expression for this specific solution?" If you can state the result immediately, you are not a beginner. I address myself, there-fore, to the beginners, and say: "Putting $y = y_0$ on the left side and $x = 0$ on the right side of Eq. (7.25) naively will get you nowhere. Take your time and think about it for a while."

Those who hesitate a little, might find it helpful to tackle the concrete problem posed at the end of the last section. In other words, set $f(x) = x^2$ and solve the problem at hand, which means that if you did not solve the problem then, you should do it now.

The question to answer now is: Why do most students find this problem easier than finding the specific solution that satisfies Eq. (7.25) and the initial condition Eq. (7.15)?

Before answering the above question, I would like to take yet another detour and show the reader a second way of finding the specific solution, starting with Eq. (7.25), but *bypassing the general solution*. As before, multiply Eq. (7.21) by dx, and integrate over x from 0 to x. Since x is now the upper limit, it is prudent to replace the variable of integration from x to some other symbol, say x_1. Let us deal with the left hand side first:

$$\int_0^x \frac{du}{dx_1}\, dx_1 = u(x) - u(0) = u(x) - y(0) = u(x) - y_0. \quad (7.26)$$

We see then that integration of Eq. (7.14) from 0 to x gives:

$$u(x) = y_0 + \int_0^x e^{kx_1} f(x_1)\, dx_1, \quad (7.27)$$

$$\therefore \quad y = y_0 e^{-kx} + \underbrace{e^{-kx} \int_0^x e^{kx_1} f(x_1)\, dx_1}_{\text{PI of Eq. (7.14)}}. \quad (7.28)$$

The expression on the right hand side of Eq. (7.28) is a spe-cific or particular solution of Eq. (7.14), not its general solution, which is given in Eq. (7.25). Although each of these solutions has two terms, the first term in Eq. (7.28), $y_0 e^{-kx}$, cannot be called the CF, because this name is reserved for the general solution of the HODE associated with the IODE whose solution is sought, whereas $y_0 e^{-kx}$ is a particular solution to Eq. (7.16).

All that remains is to return to Eq. (7.25), and show that it too leads to the same result; we begin by rewriting the equation as

$$y = C e^{-kx} + e^{-kx} F(x) \tag{7.29}$$

where
$$F(x) = \int^{\dagger} e^{kx} f(x) \, dx. \tag{7.30}$$

We now apply the condition (7.15) by setting $y = y_0$ and $x = 0$ in Eq. (7.29), which yields $C = y_0 - F(0)$, and leads to the following expression for the specific solution:

$$y = y_0 \, e^{-kx} + e^{-kx} [F(x) - F(0)] \tag{7.31}$$

$$= y_0 \, e^{-kx} + e^{-kx} \int_0^x e^{kx_1} f(x_1) \, dx_1 \tag{7.32}$$

$$= rhs\text{-}(7.28) \tag{7.33}$$

In Eq. (7.33) and in what is to follow, rhs-$(\cdots)$ means "the right hand side of Eq. $(\cdots)$, and lhs-$(\cdots)$ will have a similar meaning.

The form of the general solution given in Eq. (7.25) becomes very easy to reproduce without any calculations if one remembers the 'map' shown in Box 7.1.

Box 7.1. Alternative routes for solving Eq. (7.14)

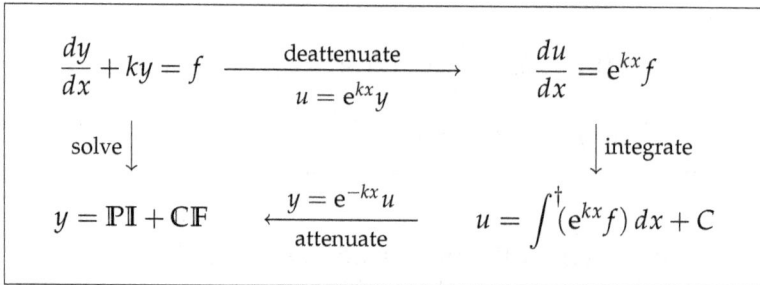

$\dfrac{dy}{dx} + ky = f$	$\xrightarrow[\;u = e^{kx}y\;]{\text{deattenuate}}$	$\dfrac{du}{dx} = e^{kx} f$
solve $\downarrow$		$\downarrow$ integrate
$y = \mathbb{PI} + \mathbb{CF}$	$\xleftarrow[\;\text{attenuate}\;]{y = e^{-kx}u}$	$u = \int^{\dagger} (e^{kx} f) \, dx + C$

deattenuate: We have gone along the route 'deattenuate, integrate, and at-
de+attenuate tenuate' to deduce the general solution. It is important to note that the task of inserting the pertinent expression for f (often called the *forcing term*) and carrying out the integration must still be completed, if the functional form of the solution is desired, which will always be our object in this book. It will be shown

later (Chapter 9) that integration can often be avoided if one follows the direct route shown in the left part of the box.

Finding the complementary function

Let us return to the statement made above that

$$y = Ae^{-kx}, \qquad (A = \text{constant}). \qquad (7.17)$$

is the general solution to Eq. (7.16), which equation will now be written as

$$\frac{dy}{dx} = -ky. \qquad (7.34)$$

Recall now that the exponential function has been defined in this book as the function that is its own derivative, which means not only that $de^s/ds = e^s$, but also that

$$\frac{d(Ae^s)}{ds} = Ae^s, \qquad (7.35)$$

where A is independent of s. The substitution $s = -kx$ converts Eq. (7.35) to Eq. (7.16). Since A can be assigned any value at will, Ae^{-kx} is the general solution to the homogeneous Eq. (7.16), and therefore the CF of Eq. (7.14).

Finding a particular integral

How one finds a particular integral is an important question that will be answered later. Since guessing has already been mentioned above, it is interesting to choose some simple expressions for the inhomogeneous term, and see if you are able to guess the right expression for $y^\dagger$. Try using $f = A_0$ and $f = A_1 x$, where A_0 and A_1 are constants.

$f = A_0$ **and** $f = A_1 x$

For the case $f = A_0$, it is not too difficult to see that $y^\dagger$ must itself be a constant, and that one must have $ky^\dagger = A_0$, or $y^\dagger = A_0/k$.

Some reflection will show that, for the case $f = A_1 x$, $y^\dagger = a_0 + a_1 x$ must be an appropriate choice. Whether it is indeed an appropriate choice, and if it is, what values should be assigned to the constants a_0 and a_1 are questions that can be easily answered by substituting the

104

guessed solution in Eq. (7.14) and examining the implications, which are shown below:

left hand side: $a_1 + k(a_0 + a_1 x) = (a_1 + ka_0) + ka_1 x$

right hand side: $A_1 t \Rightarrow ka_1 = A_1, \ a_1 + ka_0 = 0$

$$\therefore \quad y^\dagger = \frac{A_1}{k}\left(x - \frac{1}{k}\right). \tag{7.36}$$

A sinusoidal forcing term

How easy (or difficult) is it to guess the form of a PI for Eq. (7.14) when $f = \sin \omega x$?

If I had raised this question during a lecture, I would have paused and waited for one of the students to make a guess. Needless to say, I would not have chosen an unpropitious moment— for example, the first lecture on a Monday morning, when only a few have the resolve and the physical strength to get to the auditorium, and even fewer the presence of mind without which only someone suited for a career in politics or Hollywood can suppress an absent expression—in short, an occasion when no one rises to the occasion and raises his voice. As it happens, an author never faces such a mortifying experience, and I encourage you to make a guess, and see whether it satisfies the equation

$$\frac{dy}{dx} + ky = \sin \omega x. \tag{7.37}$$

If you come up with the suggestion, $y^\dagger = A \sin \omega x$, I will have to say, "It is a good first guess, but it won't do", and if you respond with the remark that this would be a good approximation if k is very large, and went on to add quickly that "$y^\dagger = B \cos \omega x$ would be a good approximation if k is very small", you would jump to the right guess before I have had the chance to say, "Congratulations, my dear fellow, you are beginning to read the equation, and you deserve a round of applause".

On the basis of what has been said above, $y^\dagger = A \sin \omega x + B \cos \omega x$ seems to be an appropriate choice. Substitution into Eq. (7.37) leads to the following relations and conclusion:

left hand side: $\sin \omega x(kA - \omega B) + \cos \omega x(\omega A + kB)$

right hand side: $\sin \omega x \Rightarrow kA - \omega B = 1, \quad \omega A + kB = 0$

$$\therefore \ y^\dagger = (k^2 + \omega^2)^{-1/2} \sin(\omega x - \delta), \text{ where } \tan \delta = \omega/k. \tag{7.38}$$

An exponential forcing term: $f = \exp(\pm i\omega x)$

We have assumed so far that we are dealing with real variables. A reader who looks at the equation

$$\frac{dy}{dx} + ky = e^{\pm i\omega x}, \tag{7.39}$$

will assume, therefore, that either the dependent variable y is a complex quantity, or that only one part (real or imaginary) of the solution is to be retained, the choice depending on whether the actual inhomogeneous term was $\cos \omega x$ or $\sin \omega x$. Let us now try to guess the PI for Eq. (7.39), and it is easy to see that

$$y^\dagger = \frac{e^{\pm i\omega x}}{k \pm i\omega} = \frac{\pm e^{i\omega x}}{R\,e^{i\vartheta}} = R^{-1}e^{\pm i(\omega x - \vartheta)} \tag{7.40}$$

where

$$R = \sqrt{k^2 + \omega^2}, \quad \tan\vartheta = \omega/k \tag{7.41}$$

An exponential forcing term: $f = \exp(-mx)$

We have succeeded in finding particular integrals by putting various guesses in the ODE itself and making whatever changes were necessary for making the left hand side equal the inhomogeneous term. The same procedure would lead us without any further work to the conclusion that

$$y^\dagger = \frac{e^{-mx}}{k - m}, \quad (k \neq m) \tag{7.42a}$$

is a particular integral of

$$\frac{dy}{dx} + ky = e^{-mx}, \tag{7.42b}$$

but answering one question often triggers another thought or two, and I would not be surprised if a reader becomes curious to find, by guessing or by manipulating Eq. (7.42a), a particular integral (to be denoted by $y_k^\dagger$) for the IODE

$$\frac{dy}{dx} + ky = e^{-kx}, \tag{7.43}$$

whose inhomogeneous term (e^{-kx}) happens to coincide (apart from a multiplicative factor, which is of no consequence in the present context) with the CF itself.

A little persistence may lead some readers to the right guess through trial and error, and I am glad to have encouraged some of you to make a little discovery of your own. Others may be tempted to retrace the steps taken in arriving at the solution, but I would remind them that in this section we are pretending that we do not know the trick whereby Eq. (7.14) can be made to yield its general solution, which includes the recipe for finding a PI.

Let us return at this point to Eq. (7.42b), and recall that we have always been speaking of *a* PI, and that two PI's can differ by a multiple of the CF. This means that, even if we add a multiple of e^{-kx} to the right hand side of Eq. (7.42a), the sum

$$y_m^\dagger = A e^{-kx} + \frac{e^{-mx}}{k-m}, \quad (k \neq m) \tag{7.44}$$

would be a PI of Eq. (7.42b). Our plan is to mould the right hand side of Eq. (7.44) into an expression that takes the indeterminate form $0/0$ when $m = k$, and to identify $y_k^\dagger$ with $\lim_{m \to k} y_m^\dagger$. All we need now is a brief pause in which we would consider the following limit

$$\lim_{m \to k} \frac{e^{-mx} - e^{-kx}}{m - k} \equiv \frac{d e^{-kx}}{dk} = -x e^{-kx}. \tag{7.45}$$

Clearly, we should put $A = 1/(m-k)$ in Eq. (7.44) and write

$$y_m^\dagger = \frac{e^{-mx} - e^{-kx}}{k - m}. \tag{7.46}$$

By combining Eq. (7.45) and Eq. (7.46) we are led to the conclusion that

$$y_k^\dagger = x e^{-kx} \tag{7.47}$$

is a PI of the IODE

$$\frac{dy}{dx} + ky = e^{-kx}. \tag{7.43}$$

Now that we have found a PI for the special case $m = k$, we go back to Eq. (7.25) to find the recipe for calculating a PI, and see that the result

$$e^{-kx} \int^\dagger e^{kx} f(x)\, dx \xrightarrow[f = e^{-kx}]{} x e^{-kx} \tag{7.48}$$

corroborates the conclusion reached above through a limiting procedure.

A different IVP

If we change the initial condition from that given in Eq. (7.15) to that specified below

$$y = y_a \quad \text{when} \quad x = a, \tag{7.49}$$

the solution of the ODE

$$\frac{dy}{dx} + ky = f \tag{7.14}$$

comes out to be

$$y = y_a e^{-k(x-a)} + e^{-kx} \int_a^x e^{kx_1} f(x_1) \, dx_1. \tag{7.50}$$

Chapter 8

Differential Equations. Part II

> The language of mathematics is full of conventions, fictions one might almost say, which have been adopted into the language for the sake of formal symmetry and conciseness of expression. A quadratic equation has generally two roots, but sometimes only one; in the latter case we say that it has 'two equal' roots. Lines meet unless they are parallel; in the latter case we say that they 'meet at infinity'—and so on. This is all pure convention, as every mathematician knows quite well.
>
> G. H. Hardy [124, p. 864]

$\mathbf{I}$N this chapter we will consider equations of the form

$$\frac{d^2y}{dx^2} + 2b\frac{dy}{dx} + c^2y = X \tag{8.1}$$

where b and c are constants and X is a function of x alone (or a constant). The central idea of this chapter is to show that anyone who knows how to solve the first order ODE

$$\frac{dy}{dx} + ky = X, \tag{8.2}$$

and fully understands the reasons for writing $2b$ instead of b and c^2 instead of c, can easily solve a second order ODE with constant coefficients. This means, of course, that we will be making frequent use of the results established in Chapter 7.

Since the expression "linear, second-order, inhomogeneous, ordinary differential equations with constant coefficients" is inconveniently long, and since we are not going to discuss differential equations of any other type, let us use the abbreviations

IODE[2] and HODE[2] (for the inhomogeneous and homogeneous versions, respectively), and agree to drop the superscript when the abbreviation is applied to a first-order ODE.

Preliminaries

As stated earlier, a differential equation by itself is not a complete statement of the problem from our point of view. To make a solution applicable to a given data set, additional information, by way of initial or boundary conditions, is necessary. The general solution of Eq. (8.1) will contain two undetermined constants, whose values can be fixed only if we have two additional pieces of information. If we are given the value of y and $y'(= Dy = dy/dx)$ at a single point, we are dealing with what is called a one-point initial value problem. If, on the other hand, the given data refer to two different values of x, as shown in Eqs. (8.3), the problem is called a two-point boundary value problem.

$$y(x_0) = y_0 \qquad\qquad y(x_1) = y_1 \qquad\qquad (8.3a)$$
$$y(x_0) = y_0 \qquad\qquad y'(x_1) = v_1 \qquad\qquad (8.3b)$$
$$y'(x_0) = v_0 \qquad\qquad y(x_1) = y_1 \qquad\qquad (8.3c)$$
$$y'(x_0) = v_0 \qquad\qquad y'(x_1) = v_1 \qquad\qquad (8.3d)$$

Standard recipe for finding the complementary function

The general solution to Eq. (8.1) can be expressed as a sum of the complementary function (CF), which is a solution of the homogeneous equation

$$\frac{d^2u}{dx^2} + 2b\frac{du}{dx} + c^2u = 0, \qquad\qquad (8.4)$$

and a particular solution (PI), which is any solution to Eq. (8.1). The CF is usually found by substituting $y = Ae^{mx}$, where A and m do not depend on x, into Eq. (8.4), and noticing that this trial solution will satisfy Eq. (8.4) only if m is a root of the equation

$$m^2 + 2bm + c^2 = 0, \qquad\qquad (8.5)$$

which will happen if

$$m = \begin{cases} -b \pm \sqrt{b^2 - c^2} = -b \pm \Delta & \text{if } b > c, \\ -b \pm \imath\sqrt{c^2 - b^2} = -b \pm \imath\Omega & \text{if } b < c. \end{cases} \tag{8.6}$$

The CF can therefore be stated as

$$u = \begin{cases} e^{-bx}[C_1 e^{\Delta x} + C_2 e^{-\Delta x}] & \text{if } b > c, \\ e^{-bx}[A \sin \Omega x + B \cos \Omega x] & \text{if } b < c. \end{cases} \tag{8.7}$$

It is worth noting that the introduction of an undamping (or deattenuating) factor through the substitution

$$z = e^{bx} y \tag{8.8}$$

eliminates the damping term $2b(dy/dx)$ from Eq. (8.4), and z satisfies the equation

$$\frac{d^2 z}{dx^2} + \Omega^2 z = 0 = \frac{d^2 z}{dx^2} - \Delta^2 z. \tag{8.9}$$

If $b = c$ (that is, $\Delta = \Omega$), we will have

$$z = A_1 + A_2 x, \tag{8.10}$$

and the CF of Eq. (8.4) takes the form shown below:

$$u = e^{-bx}(A_1 + A_2 x). \tag{8.11}$$

Finding a PI of a IODE^2 is a tedious task, except when the inhomogeneous term (the function on the right hand side) has a particularly friendly form. Most authors present three strategies and illustrate them by making a few simple choices for X. Two of these methods, known as the method of undetermined coefficients and variation of parameters, will not be discussed here; the third, based on a symbolic approach that is sometimes called "the D-operator method", will be treated in the next chapter. In this chapter, we will rely on using a substitution that reduces a IODE^2 into two IODE's, and is applicable to any choice of X.

Method I: the shortest path

We begin by choosing two (unequal) constants α_1 and α_2 which are such that

$$\alpha_1 + \alpha_2 = -2b, \qquad \alpha_1\alpha_2 = c^2, \tag{8.12}$$

which means that α_1 and α_2 are the two roots of the equation $r^2 + 2br + c^2 = 0$, or

$$r = -b \pm \epsilon, \quad \text{with} \quad \epsilon \equiv \sqrt{b^2 - c^2}. \tag{8.13}$$

Next, we introduce two functions Y_1 and Y_2 by stating the equations which they are required to fulfill:

$$\frac{dy}{dx} - \alpha_2 y = Y_1, \tag{8.14a}$$

and

$$\frac{dy}{dx} - \alpha_1 y = Y_2, \tag{8.14b}$$

We observe, first, that the required solution y can now be expressed as

$$y = \frac{Y_1 - Y_2}{\alpha_1 - \alpha_2}, \tag{8.15}$$

and, next, that Y_k ($k = 1$ or 2) satisfies a first-order ODE, because

$$\frac{dY_k}{dx} - \alpha_k Y_k = \frac{d^2y}{dx^2} - (\alpha_1 + \alpha_2)\frac{dy}{dx} + \alpha_1\alpha_2 y \tag{8.16a}$$

$$= \frac{d^2y}{dx^2} + 2b\frac{dy}{dx} + c^2 y. \tag{8.16b}$$

$$\therefore \frac{dY_k}{dx} - \alpha_k Y_k = X. \tag{8.16c}$$

The solution to Eq. (8.16c) can be immediately stated:

$$Y_k = C_k e^{\alpha_k x} + e^{\alpha_k x} F(x, \alpha_k), \tag{8.17}$$

where

$$F(x, \alpha) \equiv \int^t e^{-\alpha x} X \, dx. \tag{8.18}$$

The expression for y can now be stated:

$$y = \frac{C_1\,e^{\alpha_1 x} - C_2\,e^{\alpha_2 x}}{\alpha_1 - \alpha_2} + $$

$$\underbrace{\text{CF}}$$

$$\underbrace{\frac{1}{\alpha_1 - \alpha_2}\left[e^{\alpha_1 x}\int^\dagger e^{-\alpha_1 x}X\,dx - e^{\alpha_2 x}\int^\dagger e^{-\alpha_2 x}X\,dx\right].}_{\text{PI}} \qquad (8.19)$$

By introducing two substitutions, we have managed to transform a second-order inhomogeneous ODE into two first-order ODE's, whose solution is already known to us. All that remains is to consider the case $\epsilon = 0$ ($\alpha_1 = \alpha_2 = b$). It will be expedient to look at the PI first, write $\alpha_1 = \alpha + h$ and $\alpha_2 = \alpha - h$, where $\alpha = -b$ and $|h| = |\epsilon|$, and denote the PI by $Y_h^\dagger$. We see from Eq. (8.19) that when $h = 0$, the expression for the PI becomes a vanishing fraction. It is reasonable therefore to suppose that $Y_0^\dagger$, the required expression for the case $h = 0$, can be identified with the limiting value of the vanishing fraction; that is, $Y_0^\dagger \equiv \lim_{h\to 0} Y_h^\dagger$.

In terms of the symbols introduced below,

$$Y(x,\alpha) = e^{\alpha x}\int^\dagger e^{-\alpha x}X\,dx = e^{\alpha x}F(x,\alpha), \qquad (8.20)$$

we can write

$$\lim_{h\to 0} Y_h^\dagger = \frac{\partial Y(x,\alpha)}{\partial \alpha}. \qquad (8.21)$$

One sees that

$$Y_0^\dagger \equiv \frac{\partial Y}{\partial \alpha} = xe^{\alpha x}F(x,b) + e^{\alpha x}\frac{\partial F}{\partial \alpha}$$

$$= xe^{\alpha x}\int e^{-\alpha x}X\,dx + e^{\alpha x}\frac{\partial}{\partial \alpha}\int e^{-\alpha x}X\,dx$$

$$= xe^{\alpha x}\int e^{-\alpha x}X\,dx + e^{\alpha x}\int \frac{\partial}{\partial \alpha}e^{-\alpha x}X\,dx$$

$$= xe^{\alpha x}\int e^{-\alpha x}X\,dx - e^{\alpha x}\int x e^{-\alpha x}X\,dx$$

$$= e^{\alpha x}\left[x\int e^{\alpha x}X\,dx - \int x e^{\alpha x}X\,dx\right]$$

$$= e^{\alpha x}\left[xF - \int x\frac{\partial F}{\partial x}\,dx\right] \qquad (8.22)$$

$$= e^{\alpha x} \int F(x, \alpha)\, dx. \tag{8.23}$$

It is easily verified that $Y_0^\dagger$ satisfies the equation

$$\frac{d^2 Y_0^\dagger}{dx^2} - 2\alpha \frac{dY_0^\dagger}{dx} + \alpha^2 Y_0^\dagger = X. \tag{8.24}$$

Let us denote the CF in Eq. (8.19) by $Y_h^\bullet$. It remains to be demonstrated that $Y_0^\bullet$, the CF for the case $h = 0$, can be extracted by applying a limiting procedure to $Y_0^\bullet$. This is indeed possible, but as we already know the answer [see Eq. (8.11)], the task of taking the limit $\lim_{h \to 0} Y_h^\bullet$ is left to the reader.

Therefore, when the roots are equal ($\alpha_k = \alpha = -b$), the complete solution of Eq. (8.1) is

$$y = Y_0^\bullet + Y_0^\dagger \tag{8.25}$$

$$= e^{\alpha x}(A + Bx) + e^{\alpha x} \int \left[\int e^{-\alpha x} X\, dx \right] dx. \tag{8.26}$$

It seems worthwhile to present a technique that is applicable to equations of higher order. We will illustrate it by solving a second-order equation, and its extension to equations of higher order will be discussed in Chapter 9.

Method II: a slightly longer path

The purpose of this section is to find a solution without exploiting the symmetry of the equation. This means that we will use only Y_1 or Y_2 (the procedure being indifferent to which one we use). We choose Y_1 and write down the solution of Eq. (8.14a) as

$$y = C_2 e^{\alpha_2 x} + e^{\alpha_2 x} \int e^{-\alpha_2 x} Y_1\, dx. \tag{8.27}$$

We put $k = 1$ in Eq. (8.17) to acquire the expression for Y_1 and substitute it into Eq. (8.27) to obtain

$$y = C_2 e^{\alpha_2 x} + e^{\alpha_2 x} \int e^{\alpha_2 x} \left[C_1 e^{\alpha_1 x} + e^{\alpha_1 x} \int e^{-\alpha_1 x} X\, dx \right] dx$$

$$= \underbrace{C_2 e^{\alpha_2 x} + e^{\alpha_2 x} \left[C_1 \int e^{(\alpha_1 - \alpha_2)x}\, dx \right]}_{\text{CF}} +$$

$$e^{\alpha_2 x} \int e^{(\alpha_1 - \alpha_2)x} \left[\int e^{-\alpha_1 x} X dx \right] dx . \qquad (8.28)$$

$$\underbrace{\phantom{e^{\alpha_2 x} \int e^{(\alpha_1 - \alpha_2)x} \left[\int e^{-\alpha_1 x} X dx \right] dx}}_{\text{PI}}$$

The complementary function for this case, namely

$$\text{CF} = C_2 e^{\alpha_2 x} + \frac{C_1 e^{\alpha_1 x}}{\alpha_1 - \alpha_2}, \qquad (8.29)$$

has two undetermined constants, which is to be expected, since we have carried out two indefinite integrations. By the same token, the expression for the particular integral turns out to be

$$\text{PI} = e^{\alpha_2 x} \underbrace{\int e^{(\alpha_1 - \alpha_2)x} \left[\int e^{-\alpha_1 x} X dx \right] dx}_{=I(x,\alpha_1,\alpha_2)} \qquad (8.30)$$

$$= e^{\alpha_2 x} I(x, \alpha_1, \alpha_2), \qquad (8.31)$$

$$\text{where} \quad I(x, \alpha_1, \alpha_2) = \int e^{(\alpha_1 - \alpha_2)x} \left[\int e^{-\alpha_1 x} X dx \right] dx . \qquad (8.32)$$

On integrating $I(x, \alpha_1, \alpha_2)$ by parts, we get:

$$I(x, \alpha_1, \alpha_2) = \frac{e^{(\alpha_1 - \alpha_2)x}}{\alpha_1 - \alpha_2} \int e^{-\alpha_1 x} X dx - \frac{1}{\alpha_1 - \alpha_2} \int e^{-\alpha_2 x} X dx .$$

$$\therefore \text{PI} = \frac{1}{\alpha_1 - \alpha_2} \left[e^{\alpha_1 x} \int e^{-\alpha_1 x} X dx - e^{\alpha_2 x} \int e^{-\alpha_2 x} X dx \right], \qquad (8.33)$$

which is in agreement with Eq. (8.19).

To retrieve the solution for $\epsilon = 0$, we return to Eq. (8.28), and set $\alpha_1 = b = \alpha_2$, which gives

$$y = \underbrace{C_2 e^{-bx} + C_1 x e^{-bx}}_{\text{CF}} + \underbrace{e^{-bx} \int \left[\int e^b X dx \right] dx}_{\text{PI}} . \qquad (8.34)$$

Integration by parts now gives

$$\text{PI} = x \int e^b X dx - \int x e^b X dx . \qquad (8.35)$$

The general solution of Eq. (8.1) contains two constants whose values must be determined by imposing two supplementary conditions. For a two-point boundary value problem, this determination is a straightforward but tedious task. However, for a one-point initial value problem, the two conditions can be incorporated in the formal solution, as is shown in the following section.

A painless prescription for solving initial value problems

It was pointed out in § 7 that if the object is solve

$$\frac{dy}{dx} + ky = f(x), \tag{7.14}$$

subject to the initial condition

$$y = y_0 \quad \text{when} \quad x = 0, \tag{8.36}$$

finding this solution directly is easier than finding first the general solution and then adjusting the value of the constant of integration by imposing the initial condition on the general solution. The difference between the two approaches becomes particularly noticeable when one solves the second order differential equation, as will be shown below.

Since one-point initial value problems are usually those where time t is the independent variable, we introduce the notation

$$\dot{y}(t) \equiv \frac{dy}{dt}, \qquad \ddot{y}(t) \equiv \frac{d^2y}{dt^2}, \tag{8.37}$$

and concern ourselves with solving the equation

$$\ddot{y}(t) + 2\gamma\dot{y}(t) + w^2 y(t) = f(t), \tag{8.38}$$

subject to the initial conditions

$$\dot{y} = v_0 \quad \text{and} \quad y = u_0 \quad \text{when} \quad t = t_0. \tag{8.39}$$

Introducing now two functions $Y_1(t)$ and $Y_2(t)$,

$$Y_1(t) \equiv \frac{dy}{dt} - \lambda_2 y, \tag{8.40a}$$

$$Y_2(t) \equiv \frac{dy}{dt} - \lambda_1 y, \tag{8.40b}$$

where the constants λ_1 and λ_2 satisfy the relations $\lambda_1 + \lambda_2 = -2\gamma$ and $\lambda_1\lambda_2 = w^2$, or equivalently, the equation

$$\lambda^2 + 2\gamma\lambda + w^2 = 0. \tag{8.41}$$

Assuming that the damping is less than critical (that is, $\omega^2 > \gamma^2$), λ_1 and λ_2 come out to be

$$\lambda_1 = -\gamma + \imath\Omega, \qquad \lambda_2 = -\gamma - \imath\Omega, \qquad \left[\Omega \equiv \sqrt{\omega^2 - \gamma^2}\,\right]. \quad (8.42)$$

It follows from Eqs. (8.40) that $y(t)$ can be expressed as

$$y(t) = \frac{Y_2(t) - Y_1(t)}{\lambda_1 - \lambda_2}, \quad (8.43)$$

and that
$$\frac{dY_i}{dt} - \lambda_i Y_i = \frac{d^2y}{dt^2} - (\lambda_1 + \lambda_2)\frac{dy}{dt} + \lambda_1\lambda_2 y$$

$$\therefore \frac{dY_i}{dt} - \lambda_i Y_i = f(t). \quad (8.44)$$

The solution of Eq. (8.44) can be written as

$$Y_i(t) = e^{\lambda_i t} F(t; \lambda_i) + e^{\lambda_i(t-t_0)} Y_i(t_0), \quad (8.45)$$

where
$$F(t; \lambda_i) \equiv \int_{t_0}^{t} e^{-\lambda_i u} f(u)\, du. \quad (8.46)$$

The initial values $Y_i(t_0)$ $(i = 1,2)$ entering Eq. (8.45) can be specified by turning to Eqs. (8.40a) and (8.40b), which show that

$$Y_1(t_0) = \dot{y}(t_0) - \lambda_2 y(t_0) = v_0 - \lambda_2 u_0, \quad (8.47)$$

$$Y_2(t_0) = \dot{y}(t_0) - \lambda_1 y(t_0) = v_0 - \lambda_1 u_0. \quad (8.48)$$

$$\therefore y = \frac{1}{(\lambda_1 - \lambda_2)}\left[(v_0 - \lambda_2 u_0)\, e^{\lambda_1 t} - (v_0 - \lambda_1 u_0)\, e^{\lambda_2 t} + \right.$$

$$\left. e^{\lambda_1 t}\int_{t_0}^{t} e^{-\lambda_1 u} f(u)\, du - e^{\lambda_2 t}\int_{t_0}^{t} e^{-\lambda_2 u} f(u)\, du \right]. \quad (8.49)$$

We have found, not the general solution to the generic IODE^2, namely Eq. (8.38), but a particular solution that satisfies two given initial conditions, and this is sufficient for anyone who is looking for the solution to a given initial value problem.

Heaviside's comment on arbitrary constants

In the light of the above derivation, let us look at what Oliver Heaviside wrote in a similar context [125, p. 392].

"In the usual treatment of the theory of differential equations, there is no such definiteness. Arbitrary constants are brought in to any extent, to be afterwards got rid of. Now this is all very well in the general theory of differential equations, where arbitrary constants form a part of the theory itself, but for the practical purpose of representing and obtaining solutions of physical problems, the use of such arbitrary and roundabout methods (which are too often followed, especially by elementary writers) leads to a large amount of unnecessary work, tending to obscure the subject, without helping one on. It would not, perhaps, be going too far to say that such a misuse or inefficient application of analysis often makes rigmarole."

The painful alternative

The solution stated in Eq. (8.49) suffers from one disadvantage: its components (various terms) are grouped differently from what one sees in the most common (or standard) version (which will make its appearance shortly). If an examinee is asked to reproduce the standard version, the algebraical steps involved in going from the right hand side of Eq. (8.49) to the desired goal will be the same as those in deriving the latter solution, in which case it would be better to follow the traditional route.

To present the standard version, I go back to a question posed in 1876 by Lord Rayleigh in an examination paper [126, p. 282].

Show that the solution of the differential equation for vibrations resisted by a frictional force proportional to the velocity, but otherwise free, viz

$$\ddot{u} + \kappa \dot{u} + n^2 u = 0,$$

may be put into the form

$$u = e^{-\frac{1}{2}\kappa t}\left\{\dot{u}_0\frac{\sin n't}{n'} + u_0\left(\cos n't + \frac{\kappa}{2n'}\sin n't\right)\right\}$$

where $n'^2 = n^2 - \frac{1}{4}\kappa^2$, and $\dot{u}_0$, u_0 are the values of the velocity and displacement when $t = 0$.

Deduce the complete solution of

$$\ddot{u} + \kappa\dot{u} + n^2 u = U,$$

in the form

$$u = e^{-\frac{1}{2}\kappa t}\left\{\dot{u}_0\frac{\sin n't}{n'} + u_0\left(\cos n't + \frac{\kappa}{2n'}\sin n't\right)\right\} +$$
$$\frac{1}{n'}\int_0^t e^{-\frac{1}{2}\kappa(t-t')}\sin n'(t-t')U'dt', \tag{8.50}$$

where U' is the same function of t' as U is of t.

The royal road to impulse response

A problem that frequently arises in studies of linear dynamic systems is the determination of the impulse response. The approach presented in the preceding section provides a straightforward route for deducing the impulse response, even when the right hand side contains δ-function and/or its derivative(s). No attempt will be made to substantiate this remarks, because the topic lies outside the scope of this book.

A note on operator algebra (only for the novice)

When one writes

$$\frac{dy}{dx} - ay = Dy - ay, \tag{8.51}$$

one uses a shorthand in which D stands for dy/dx. On the other hand, if one goes on to write

$$\frac{dy}{dx} - ay = Dy - ay = (D-a)y, \tag{8.52}$$

one adopts what used to be called "the method of the separation of the symbols of operation from those of quantity" (where *quantity* is to be interpreted as *function*).

Since the expression $(D-b)y$ is to be interpreted similarly, the expression $[(D-a)+(D-b)]y$ will stand for $2(dy/dx) - (a+b)y$, which means that we may write

$$[(D-a)+(D-b)]y = [2D-(a+b)]y. \tag{8.53}$$

That is, the result of operating on y with $(D-a)$ and $(D-b)$ separately, and then adding the resulting expressions is the same as operating on y with $[2D-(a+b)]$. We see that what we get by adding the results of two operations of the type considered above can be obtained by working with the sum of the two operators. Thus we can write

$$[(D-a)+(D-b)] = [2D-(a+b)]. \qquad (8.54)$$

Evidently this rule applies to a linear combination of any number of such operators.

The expression $(D-b)(D-a)y$ means

$$\left(\frac{d}{dx}-b\right)\left(\frac{dy}{dx}-ay\right) = \frac{d^2y}{dx^2} - (a+b)\frac{dy}{dx} + aby. \qquad (8.55)$$

That is, the result of operating on y with $(D-a)$ first and then with $(D-b)$ on the result of the first operation is identical with operating on y with $[D^2 - (a+b)D + ab]$. Hence we see that the operation resulting from the successive application of two operations of the type here considered, can be obtained by applying the product of their symbolic representatives. This can be expressed by writing

$$(D-b)(D-a) = D^2 - (a+b)D + ab. \qquad (8.56)$$

One sees that the interchange of a and b on the right hand side leaves the expression unaltered, which means that the order of the operators on the left is immaterial; in other words, the two operations are commutative.

The symbols for the operators of the type considered above behave like algebraic quantities for the processes of addition, subtraction, and multiplication. Though the full power of operator algebra can be appreciated only when one considers (as we shall do in Chapter 9) *inverse* operators like $(D-a)^{-1}$, factorization of a second-order operator can be used to achieve a slightly crisper exposition of the two methods labelled I and II in the preceding pages; however, steps involving operator algebra and the name *factorization method* were purposely avoided in order to make it clear that the required solution can be obtained without resorting to symbolic calculus. The analysis presented above, which relied exclusively on finding effective substitutions, is restated below in the parlance of operator algebra.

Method I

Equation (8.1) can be factorized in two different ways

$$(D - \alpha_1)\underbrace{(D - \alpha_2)y}_{=Y_1} = X = (D - \alpha_2)\underbrace{(D - \alpha_1)y}_{=Y_2}. \qquad (8.57)$$

Define

$$(D - \alpha_2)y = Y_1, \qquad (8.58a)$$
$$(D - \alpha_1)y = Y_2, \qquad (8.58b)$$

and note that subtraction of the last two equations gives

$$y = \frac{Y_1 - Y_2}{\alpha_1 - \alpha_2}, \qquad (8.59)$$

where Y_k ($k = 1, 2$) satisfies the IODE[1]

$$(D - \alpha_k)Y_k = X. \qquad (8.60)$$

Method II

Start by writing Equation (8.1) as

$$(D - \alpha_1)\underbrace{(D - \alpha_2)y}_{=Y_1} = X.$$

Find Y_1 by solving $(D - \alpha_1)Y_1 = X$, and find y by solving the equation $(D - \alpha_1)y = Y_1$

The same results would be obtained if one starts by writing Equation (8.1) as

$$(D - \alpha_2)\underbrace{(D - \alpha_1)y}_{=Y_2} = X.$$

and finds Y_2 and thereafter y.

Chapter 9

Conversion of Operational to Algebraic Forms

Perhaps some people will say (as usual) that they do not like "algebrize"; that it is un-English, &c., &c. People are always saying something. What is more important is that a word to express the idea of conversion from operational to algebraical form is much wanted, and that "algebrize" seems to answer the purpose very well. Similarly we might say that we logarize a number when we take its logarithm, and delogarize it when we find the number whose logarithm it is; and so on.

Oliver Heaviside [122, p. 41]

There is no doubt that this operational method very greatly curtails the mathematical work in a large number of cases, although it occasionally happens that the process of "algebrising" the operational form of solution involves some troublesome mathematical work of the ordinary (or humdrum) sort. Mr. Heaviside is, however, always at home with this method, while beginners will have to mind their steps; but they will be amply rewarded for any time and trouble that they take in studying the process [127].

I N Chapter 8, we saw that, so far as linear ordinary differential equations with constant coefficient are concerned, the basic building block is the solution of the the first-order, inhomogeneous ordinary differential equation (IODE)

$$\frac{dy}{dx} + ky = f(x). \tag{7.14}$$

The solution is a sum of two terms, the complementary function (CF) and a particular integral (PI):

$$y = \underbrace{C e^{-kx}}_{\text{CF}} + \underbrace{e^{-kx} \int^t e^{kx} f(x)\, dx}_{\text{PI of Eq. (7.14)}}. \tag{7.25}$$

For practical applications one prefers to express the integral on the right hand side in terms of known functions. The purpose of this chapter is to show that whenever f happens to be a student-friendly function (see below), one can find a PI more easily by algebrizing the operational expression than by going through the tedious task of integration.

Solutions of linear ODE's with constant coefficients

As stated earlier (p. 100), the general solution of the IODE

$$(D+k)y = f \tag{9.1}$$

can be expressed as $y = y^\dagger + y_h$ where y_h is the general solution of the corresponding homogeneous equation,

$$(D+k)y = 0, \tag{9.2}$$

and $y^\dagger$ is any solution to Eq. (9.1).

In view of the foregoing, we will rephrase Eq. (9.1) as

$$(D+k)y = 0 + f, \tag{9.3}$$

and express its formal solution as

$$y = \underbrace{\frac{1}{D+k}0}_{\text{CF}} + \underbrace{\frac{1}{D+k}f}_{\text{PI}}. \tag{9.4}$$

The meaning of the inverse operator

$$\frac{1}{D+k} = (D+k)^{-1} \tag{9.5}$$

is that whatever the inverse operator $(D+k)^{-1}$ does to a function is undone by the application of the direct operator $D+k$. We express this relation as follows

$$(D+k)[(D+k)^{-1}f] = f, \quad \text{or} \quad (D+k)(D+k)^{-1} = 1. \quad (9.6)$$

If we apply the operator $(D+k)$ to both sides of Eq. (9.4), we evidently recover Eqs. (9.3) and (9.1).

Since the special case $k = 0$, which corresponds to antidifferentiation, will also be considered here, we put $k = 0$ in Eq. (9.3) to get the equation

$$Dy = 0 + f, \quad (9.7)$$

and express its formal solution as

$$y = \underbrace{\frac{1}{D}0}_{C} + \underbrace{\frac{1}{D}f}_{\text{PAD}}, \quad (9.8)$$

where, we recall, C is the so-called constant of integration and the abbreviation PAD stands for *particular antiderivative* (or *particular antidifferential*, if one is dealing with a differential). If we use the vocabulary of DE's, C is the CF (being the general solution of the equation $Dy = 0$), and any PAD is a PI of the equation $Dy = f$.

Our aim in this chapter is to develop symbolic methods for finding $y^\dagger$ and y_h by treating D, $(D+k)$ and their inverses

$$D^{-1} \equiv \frac{1}{D} \quad \text{and} \quad (D+k)^{-1} \equiv \frac{1}{D+k}$$

as if they were symbols in ordinary algebra. It will be our task to give meanings to whatever inverse operators arise in our work. In what follows, it will be sufficient to speak of only the inverse operator $(D+k)^{-1}$, since it subsumes D^{-1} as a special case.

The first point to note is that, since we are dealing with a first-order ODE, the CF of Eq. (9.2) must contain one arbitrary constant (neither more nor fewer), and a PI must *not* contain an *arbitrary* constant. This means that the meaning attached to an inverse operator will be determined by what one is trying to find, the CF or a PI; in other words, the meaning will depend on whether the operand is 0 or f (a non-zero quantity, which may be a function of x or a constant).

It will be shown shortly (p. 128) that

$$\frac{1}{D+k}0 = Ce^{-kx},$$

which implies that

$$\frac{1}{D}0 = C,$$

a reassuring conclusion, for we have already interpreted D^{-1} as the operator for antidifferentiation. The results of applying D^{-2} to 0 is shown below:

$$D^{-2}0 = D^{-1}[D^{-1}0] = D^{-1}C_1 = D^{-1}(0+C_1) = C_2 + C_1x,$$

where C_1 and C_2 are arbitrary constants; as is to be expected, the result agrees with that found earlier by direct integration [cf. Eq. (7.5)].

The exponential function

We have already solved the following IVP

$$\begin{cases} \dfrac{dy}{dx} = -ky, & \text{(6.11a)} \\ y(0) = y_0 & \text{(6.11b)} \end{cases}$$

by using an iterative method. We now want to recover the same result by using operational calculus, but we consider first the related IVP stated below:

$$\begin{cases} \dfrac{dy}{dx} = y, & \text{(9.10a)} \\ y(0) = 1. & \text{(9.10b)} \end{cases}$$

We begin by transposing y to the left hand side of (9.10a) and manipulating the resulting equation:

$$\frac{dy}{dx} - y = 0 \tag{9.11}$$

$$(D-1)y = 0 \tag{9.12}$$

$$D\underbrace{\left[1 - D^{-1}\right]}_{=\,?}y = 0. \tag{9.13}$$

The vanishing of *lhs*-(9.13) implies that $[1 - D^{-1}]y$ must be independent of x. Therefore

$$\left[1 - D^{-1}\right] y = C, \quad \text{or} \quad y = \frac{C}{1 - D^{-1}}. \tag{9.14}$$

In order to interpret the denominator of the right hand side of the second equality in (9.14), we seek the aid of the formula

$$1 + a + a^2 + a^3 + \ldots = \frac{1}{1 - a}, \quad (a < 1) \tag{9.15}$$

and expand $1/(1 - D^{-1})$ likewise to see if what we obtain makes any sense. The result, shown below, makes perfect sense:

$$y = \frac{1}{1 - D^{-1}} C \tag{9.16}$$

$$= C \left[1 + D^{-1} + D^{-2} + D^{-3} + \cdots\right] 1 \tag{9.17}$$

$$= C \left[1 + x + \frac{x^2}{2} + + \frac{x^3}{3!} + \cdots\right] \tag{9.18}$$

$$= C \sum_{n=0}^{\infty} \frac{x^n}{n!} = Ce^x. \tag{9.19}$$

The infinite series can be shown to be convergent, but this task is considered unnecessary here. Upon imposing the initial condition $y(0) = 1$, we get $C = 1$, which implies that

$$y = e^x,$$

and we see that y is in fact the exponential function $\exp(x)$, being its own derivative and equalling unity when the argument is zero. The value of e can be calculated by setting $x = 1$ and retaining as many terms as yield the desired accuracy.

The complementary function

To deduce the general solution of the homogeneous equation $(D + k)y_h = 0$, we take the following steps, the rationale for which need not be given:

$$(D + k)y_h = 0 \tag{9.20}$$

$$D\left[\underbrace{(1+kD^{-1})y_h}_{=\text{constant}}\right] = 0 \qquad (9.21)$$

$$\therefore (1+kD^{-1})y_h = C \qquad (9.22)$$

The reader will appreciate that the derivation is replicating the approach used in the preceding section; accordingly, no explanation will be offered for the remaining steps, which are displayed below:

$$\begin{aligned}
y_h &= \frac{1}{1+kD^{-1}}C \\
&= [1+kD^{-1}+k^2D^{-2}+k^3D^{-3}+\ldots]C \\
&= C\left[1-kx+\frac{(kx)^2}{2}-\frac{(kx)^3}{3!}+\ldots\right] \\
&= Ce^{-kx}.
\end{aligned} \qquad (9.23)$$

Some identities

In this section, we will establish the truth of some identities that will prove to be useful later on.

The following result can be verified by direct differentiation:

$$(D+k)e^{ax} = e^{ax}(a+k). \qquad (9.24)$$

If we treat $D+k$ as an ordinary algebraic expression, we immediately arrive at the following relation

$$\frac{1}{D+k}e^{ax} = \frac{1}{a+k}e^{ax}, \qquad (k+a \neq 0). \qquad (9.25)$$

To see whether Eq. (9.25) is indeed correct, we operate on both sides of the equation by $(D+k)$, and use the golden rule that the direct operator $(D+k)$ must annul $(D+k)^{-1}$, or equivalently

$$(D+k)\frac{1}{D+k} = 1. \qquad (9.26)$$

Applying $(D+k)$ to both sides of Eq. (9.25), we find

$$(D+k)\frac{1}{D+k}e^{ax} = (D+k)\frac{1}{a+k}e^{ax} \qquad (9.27)$$

$$lhs\text{-}(9.27) = e^{ax} \tag{9.28}$$

$$rhs\text{-}(9.27) = (D+k)\frac{1}{a+k}e^{ax} = \frac{1}{a+k}\underbrace{(D+k)e^{ax}}_{=(a+k)e^{ax}} \tag{9.29}$$

$$= \frac{1}{a+k}(a+k)e^{ax} = e^{ax}, \tag{9.30}$$

which establishes Eq. (9.25), except when $a = -k$.

To find $(D+k)^{-1}e^{-kx}$, we will first look at the equation

$$De^{kx}V = e^{kx}(D+k)V, \tag{9.31}$$

also found by differentiation, and ask if its inverse counterpart, namely

$$D^{-1}e^{kx}V \overset{?}{=} e^{kx}(D+k)^{-1}V, \tag{9.32}$$

is also valid. Applying D to rhs-(9.32), we get

$$D[rhs\text{-}(9.32)] = D\big[e^{kx}\underbrace{(D+a)^{-1}V}_{=W,\,\text{say}}\big] \tag{9.33}$$

$$= D\big[e^{kx}W\big] = e^{kx}(D+k)\big[W\big] \tag{9.34}$$

$$= e^{kx}(D+k)\big[(D+k)^{-1}V\big] \tag{9.35}$$

$$= e^{kx}V = D[lhs\text{-}(9.32)]. \tag{9.36}$$

Thus, we have established the identity

$$D^{-1}e^{kx}V = e^{kx}(D+k)^{-1}V. \tag{9.37}$$

Multiplying both sides by e^{-kx}, and interchanging the two sides of the resulting equation, we get

$$\frac{1}{D+k}V = e^{-kx}D^{-1}e^{kx}V \tag{9.38}$$

$$= e^{-kx}\int^{\dagger}e^{kx}V\,dx. \tag{9.39}$$

Equation (9.38) provides another interpretation of the operation $(D+k)^{-1}$, but the expression on the right hand side of Eq. (9.39) is simply the standard recipe for finding a PI of the ODE at hand, namely $(D+k)y = V$. Our aim is to avoid using D^{-1}, except when it operates on a singularly simple function, such as a constant, x^m ($m = 0,1,2,\ldots$), e^{mx}, $\sin mx$ and $\cos mx$. If we set $V = e^{-kx}$ in Eq. (9.38), $V = e^{kx}e^{-kx} = 1$, and we get

$$\frac{1}{D+k}e^{-kx} = e^{-kx}\frac{1}{D}1 = xe^{-kx}. \tag{9.40}$$

Sinusoidal right hand side

When $X = \sin\omega x$ or $X = \cos\omega x$, the quickest way to find $(D+k)X$ is to use the complex representation and write $\cos\omega x = \Re\{e^{i\omega x}\}$ and $\sin\omega x = \Im\{e^{i\omega x}\}$, and amend Eq. (9.25) to read

$$\frac{1}{D+k}e^{i\omega x} = \frac{1}{k+i\omega}e^{i\omega x}. \tag{9.41}$$

The denominator $k + i\omega$ can be expressed as

$$k + i\omega = \rho\, e^{i\phi}, \tag{9.42}$$

where

$$\rho = \sqrt{k^2 + \omega^2}, \quad \phi = \arctan\left(\frac{\omega}{k}\right). \tag{9.43}$$

$$\Re\left\{\frac{1}{D+k}e^{i\omega x}\right\} = \frac{1}{\rho}\Re\left\{e^{i(\omega x-\phi)}\right\} = \frac{1}{\rho}\cos(\omega x - \phi) \tag{9.44}$$

$$\Im\left\{\frac{1}{D+k}e^{i\omega x}\right\} = \frac{1}{\rho}\Im\left\{e^{i(\omega x-\phi)}\right\} = \frac{1}{\rho}\sin(\omega x - \phi) \tag{9.45}$$

An alternative

An alternative method makes use of the following identities, the first two of which result from differentiation, the verification of the next two requires the application of the operator $(D^2 - \omega^2)$ to the first two, and the last is simply a result of treating D like a symbol of ordinary algebra.

$$(D^2 - k^2)\sin\omega x = -(\omega^2 + k^2)\sin\omega x \tag{9.46}$$
$$(D^2 - k^2)\cos\omega x = -(\omega^2 + k^2)\cos\omega x \tag{9.47}$$
$$(D^2 - k^2)^{-1}\sin\omega x = -(\omega^2 + k^2)^{-1}\sin\omega x \tag{9.48}$$
$$(D^2 - k^2)^{-1}\cos\omega x = -(\omega^2 + k^2)^{-1}\cos\omega x \tag{9.49}$$

$$\frac{1}{D+k} = \frac{D-k}{D^2-k^2} = (D-k)\frac{1}{D^2-k^2} \qquad (9.50)$$

We are thus led to the following calculation:

$$\frac{1}{D+k}\sin\omega x = (D-k)\frac{1}{D^2-k^2}\sin\omega x \qquad (9.51)$$

$$= (D-k)\frac{1}{-(\omega^2+k^2)}\sin\omega x \qquad (9.52)$$

$$= -\frac{1}{(\omega^2+k^2)}(D-k)\sin\omega x \qquad (9.53)$$

$$= \frac{1}{\sqrt{\omega^2+k^2}}\sin(\omega x-\phi). \qquad (9.54)$$

In going from Eq. (9.53) to Eq. (9.54), we have used the result (with the minus sign)

$$(D\pm k)\sin\omega x = \pm\sqrt{\omega^2+k^2}\sin(\omega x\pm\phi), \qquad (9.55)$$

and the plus sign has been included to show that while the application of $(D+k)$ to $y=\sin\omega x$ multiplies y by $\rho=\sqrt{\omega^2+k^2}$, and advances the phase by ϕ, the application of $(D+k)^{-1}$ to y divides it by ρ and retards the phase by ϕ [Eq. (9.54)].

A forcing term containing x^n

Unlike the previous cases, there seems to be only one way of handling a forcing term of the type $f=x^n$. To be concrete, let us take $n=3$, and denote (for reasons that will become clear presently) the the PI by $y_k^\dagger$.

$$y_k^\dagger = \frac{1}{D+k}x^3 \qquad (9.56)$$

$$= \frac{1}{k\left[1+\dfrac{D}{k}\right]}x^3 = \frac{1}{k}\left[1-\frac{D}{k}+\frac{D^2}{k^2}-\frac{D^3}{k^3}+\cdots\right]x^3$$

$$= \frac{1}{k}\left[x^3-\frac{3x^2}{k}+\frac{6x}{k^2}-\frac{6}{k^3}\right]. \qquad (9.57)$$

It is evident that, when one is dealing with $f=x^n$, the series terminates when one reaches the term containing D^n.

Evaluation of some antiderivatives

Let us recall the expression given in Eq. (7.25) for the $\mathbb{PI}$ of our 1ODE-I,

$$y_k^\dagger = e^{-kx} \int^\dagger e^{kx} f(x) \, dx, \tag{9.58}$$

and multiply by e^{kx} to obtain

$$e^{kx} y_k^\dagger = \int^\dagger e^{kx} f(x) \, dx \tag{9.59}$$

Integrals of the type appearing on the right hand side often arise in elementary calculus, with $f(x)$ given by $\sin mx$, $\cos mx$, or x^m, and it is customary to integrate them by parts. For example, the integral defined in Eq. (9.60),

$$I_n \equiv \int^\dagger e^{kx} x^n \, dx, \tag{9.60}$$

transforms, after one round of integration by parts, into

$$I_n = \frac{x^n}{k} e^{kx} - \frac{n}{k} I_{n-1}. \tag{9.61}$$

For $n = 3$, the result came out to be:

$$I_3 = \frac{e^{kx}}{k} \left[x^3 - \frac{3x^2}{k} + \frac{6x}{k^2} - \frac{6}{k^3} \right]. \tag{9.62}$$

If we refer to Eq. (9.57), we find

$$\frac{1}{D+k} x^3 = \frac{1}{k} \left[x^3 - \frac{3x^2}{k} + \frac{6x}{k^2} - \frac{6}{k^3} \right]. \tag{9.63}$$

On multiplying the expression on the right hand side by e^{kx}, we recover the result obtained in Eq. (9.62).

Two shift theorems

Consider the problem of evaluating the following $\mathbb{PI}$

$$y_k^\dagger = \frac{1}{D+k} e^{mx} V, \tag{9.64}$$

where V is some function of x. Since one factor in the forcing term $X = e^{mx}V$ is e^{mx}, it would be convenient to merge it with e^{kx} which also enters as a factor in the integrand of a PI. This is easily accomplished by setting $p = k + m$ and going along the route shown below:

$$y_k^\dagger = e^{-kx}\int^\dagger e^{kx} X\, dx = e^{-kx}\int^\dagger e^{kx} e^{mx} V\, dx \qquad (9.65)$$

$$= e^{mx} e^{-px}\int^\dagger e^{px} V\, dx \qquad (9.66)$$

$$= e^{mx}\frac{1}{D+p}V. \qquad (9.67)$$

Whence follows the desired formula:

$$\frac{1}{D+k}e^{mx}V = e^{mx}\frac{1}{D+m+k}V. \qquad (9.68)$$

If we set $p = 0$ (that is, $m = -k$), the above formula reduces to

$$\frac{1}{D+k}e^{-kx}V = e^{-kx}\frac{1}{D}V, \qquad (9.69)$$

which is another way of writing Eq. (9.37). With $V = 1$, Eq. (9.69) becomes identical with Eq. (9.40), the formula used for finding $y_k^\dagger$ when the forcing term is e^{-kx}, in which case the standard formula

$$\frac{1}{D+k}e^{ax} = \frac{1}{a+k}e^{ax} \qquad (9.70)$$

becomes inapplicable.

We will refer to Eq. (9.68) as the *shift theorem for exponentials*. In deducing this identity we relied on the definition of the PI, but it can also be proved by entirely operational means, as shown below:

$$(D+k)[lhs\text{-}(9.68)] = e^{ax}V \qquad (9.71)$$

$$(D+k)[rhs\text{-}(9.68)] = (D+k)\left[e^{ax}\underbrace{(D+k+a)^{-1}V}\right] \qquad (9.72)$$
$$\qquad\qquad =W,\text{ say.}$$

$$= (D+k)\left[e^{ax}W\right] = e^{ax}(D+k+a)[W] \qquad (9.73)$$

$$= e^{ax}(D+k+a)\left[(D+k+a)^{-1}V\right] \tag{9.74}$$

$$= e^{ax}V = \text{rhs-(9.71)}. \tag{9.75}$$

Before presenting the next shift theorem, let us try to find a PI for the 1ODE-I

$$\frac{dy}{dx} - y = x \sin x \tag{9.76}$$

The standard procedure leads to the following expression for the general solution:

$$y = Ce^x + e^x \int^t e^{-x} x \sin x \, dx. \tag{9.77}$$

To find a PI by operational means, we will use the formula

$$\frac{1}{D+k}xV = x\frac{1}{D+k}V - \frac{1}{(D+k)^2}V, \tag{9.78}$$

where V is some function of x; a proof will be provided shortly.
  Using the identity in Eq. (9.78), we get

$$\mathbb{PI} = \frac{1}{D-1}x \sin x = x\frac{1}{D-1}\sin x - \frac{1}{(D-1)^2}\sin x \tag{9.79}$$

$$= x\frac{D+1}{D^2-1}\sin x - \frac{D^2+1}{(D^2-1)^2}\sin x \tag{9.80}$$

$$= -\tfrac{1}{2}(\cos x + \sin x) - \tfrac{1}{2}\cos x. \tag{9.81}$$

Proof of Eq. (9.78):

As a preliminary, we introduce the symbol

$$\frac{1}{D+k}V = W, \quad \text{so that} \quad (D+k)W = V \tag{9.82}$$

and note that

$$(D+k)xW = x(D+k)W + W = xV + W \tag{9.83}$$
$$(D+k)^2xW = (D+k)(xV+W) \tag{9.84}$$
$$= (D+k)xV + V. \tag{9.85}$$

We will now show that $(D+k)^2[\text{lhs-(9.78)}] = (D+k)^2[\text{rhs-(9.78)}]$:

$$(D+k)^2[\text{lhs-(9.78)}] = (D+k)xV \tag{9.86}$$

$$(D+k)^2[\text{rhs-(9.78)}] = (D+k)^2\Big[x\underbrace{\frac{1}{D+k}V}_{=W}\Big] - V$$

$$= (D+k)xV. \tag{9.87}$$

We may call Eq. (9.78) the shift theorem for x. We have proved the theorem, but not asked how anyone may be led to it naturally. As a preamble to answering this question, it will be well to interpret the expression $(D+k)^{-2}X$:

$$\frac{1}{D+k}X = e^{-kx}\int^\dagger e^{kx}X\,dx \tag{9.88}$$

$$\frac{1}{(D+k)^2}X = e^{-kx}\int^\dagger e^{kx}\frac{1}{D+k}X\,dx \tag{9.89}$$

$$= e^{-kx}\int^\dagger e^{kx}\Big[e^{-kx}\int e^{kx}X\,dx\Big]dx \tag{9.90}$$

$$= e^{-kx}\int^\dagger\Big[\int e^{kx}X\,dx\Big]dx \tag{9.91}$$

$$= e^{-kx}\int^\dagger\int^\dagger e^{kx}X\,(dx)^2. \tag{9.92}$$

We now return to Eq. (9.88), choose $X = xV$, and integrate by parts:

$$y^\dagger = \frac{1}{D+k}xV = e^{-kx}\int^\dagger e^{kx}xV\,dx \tag{9.93}$$

$$= e^{-kx}\Big[x\int^\dagger e^{kx}V\,dx - \int^\dagger\Big[\int^\dagger e^{kx}V\,dx\Big]dx\Big] \tag{9.94}$$

$$y = xe^{-kx}\int^\dagger e^{kx}V\,dx - e^{-kx}\int^\dagger\Big[\int^\dagger e^{kx}V\,dx\Big]dx \tag{9.95}$$

$$= x\frac{1}{D+k}V - \frac{1}{(D+k)^2}V. \tag{9.96}$$

For $k = 0$, this reduces to

$$\frac{1}{D}xV = x\frac{1}{D}V - \frac{1}{D^2}V, \tag{9.97}$$

which is the operator equivalent of

$$\int^\dagger xV\,dx = x\int^\dagger V\,dx - \int^\dagger \left[\int^\dagger V\,dx\right]dx, \tag{9.98}$$

and it can be thrown in the familiar form

$$\int^\dagger x\,dW = xW - \int^\dagger W\,dx$$

by setting $dW = V\,dx$.

Second-order equations

A particular integral of the IODE[2] stated below

$$(D^2 + 2bD + c^2)y = X \tag{9.99}$$

can be found by using one of the following equations:

$$y^\dagger = \frac{1}{D^2 + 2bD + c^2}X \tag{9.100}$$

$$y^\dagger = \frac{1}{(D-\alpha)}\left[\frac{1}{(D-\beta)}X\right]$$
$$= \frac{1}{(D-\beta)}\left[\frac{1}{(D-\alpha)}X\right] \qquad (Method\text{-}2)$$

$$y^\dagger = \frac{1}{\alpha-\beta}\left[\frac{1}{D-\alpha} - \frac{1}{D-\beta}\right]X$$
$$= \frac{1}{\alpha-\beta}\left[\frac{1}{D-\alpha}X - \frac{1}{D-\beta}X\right] \qquad (Method\text{-}1)$$

Since the aim of this book is to use only operators of the type $(D+k)^{-1}$, we will only consider Eqs. (*Method-2*) and (*Method-1*), which are the operator analogs of Method II and Method I (respectively), described at the end of Chapter 8 (p. 121).

Some examples

1. $$\frac{dy}{dx} - 2y = \cos 2x$$

We have

$$(D-2)y = \cos 2x \tag{9.101}$$

$$\therefore y^\dagger = \frac{1}{D-2}\cos 2x \tag{9.102}$$

$$= (D+2)\frac{\cos 2x}{D^2-4} = (D+2)\frac{\cos 2x}{-4-4} \tag{9.103}$$

$$= -\tfrac{1}{8}(-2\sin 2x + 2\cos 2x) = \tfrac{1}{4}(\sin 2x - \cos 2x)$$

2. $$\frac{d^2y}{dx^2} + 4y = \sin 2x$$

We have

$$(D^2+4)y = \sin 2x \tag{9.104}$$

$$\therefore y^\dagger = \frac{1}{D^2+4}\sin 2x \tag{9.105}$$

$$y^\dagger = \Im\frac{1}{D-2\imath}\frac{1}{D+2\imath}e^{2\imath x} = \Im\frac{1}{D-2\imath}\frac{1}{4\imath}e^{2\imath x} \tag{9.106}$$

$$= \Im\frac{1}{4\imath}\frac{1}{D-2\imath}e^{2\imath x} = \Im\frac{1}{4\imath}e^{2\imath x}\frac{1}{D}1 \tag{9.107}$$

$$= \Im\{-\frac{\imath}{4}xe^{2\imath x}\} = -\frac{1}{4}x\cos 2x \tag{9.108}$$

Alternatively, we can write

$$y^\dagger = \Im\frac{1}{D+2\imath}\frac{1}{D-2\imath}e^{2\imath x} = \Im\frac{1}{D+2\imath}e^{2\imath x}\frac{1}{D}1 \tag{9.109}$$

$$= \Im e^{2\imath x}\frac{1}{D+4\imath}x = \Im e^{2\imath x}\frac{1}{4\imath\left(1+\frac{D}{4\imath}\right)}x \tag{9.110}$$

$$= \Im\frac{1}{4\imath}e^{2\imath x}\left(1-\frac{D}{4\imath}+\cdots\right)x = \Im\frac{1}{4\imath}e^{2\imath x}x \tag{9.111}$$

3. $$\frac{d^2y}{dx^2} + 2b\frac{dy}{dx} + c^2y = \sin \omega x$$

We have

$$y^\dagger = \frac{1}{D^2+2bD+c^2}\sin \omega x$$

$$= \frac{D^2 + c^2 - 2bD}{D^2 + c^2 - 2bD} \cdot \frac{1}{D^2 + c^2 + 2bD} \sin \omega x$$

$$= (D^2 + c^2 - 2bD) \frac{1}{(D^2 + c^2)^2 - 4b^2 D^2} \sin \omega x$$

$$= (D^2 + c^2 - 2bD) \frac{1}{(-\omega^2 + c^2)^2 + 4b^2 \omega^2} \sin \omega x$$

$$= (c^2 - \omega^2 - 2bD) \frac{1}{(c^2 - \omega^2)^2 + 4b^2 \omega^2} \sin \omega x$$

$$= \frac{(c^2 - \omega^2) \sin \omega x - 2b\omega \cos \omega x}{(c^2 - \omega^2)^2 + 4b^2 \omega^2}$$

$$= \frac{\sin(\omega x - \phi)}{\sqrt{(c^2 - \omega^2)^2 + 4b^2 \omega^2}} \tag{9.112}$$

where
$$\tan \phi = \frac{2b\omega}{c^2 - \omega^2}. \tag{9.113}$$

One can also use factorization and follow *Method-2* by going along the path shown below.

$$y^\dagger = \Im \frac{1}{(D + \lambda_1)(D + \lambda_2)} e^{\imath \omega x} \tag{9.114}$$

$$= \Im \frac{1}{(D + \lambda_1)} \left[\frac{1}{(D + \lambda_2)} e^{\imath \omega x} \right] = \Im \frac{1}{(D + \lambda_1)} \frac{1}{(\imath \omega t + \lambda_2)} e^{\imath \omega x}$$

$$= \Im \frac{1}{(\lambda_2 + \imath \omega t)} \left[\frac{1}{(D + \lambda_1)} e^{\imath \omega x} \right] = \Im \frac{1}{(\lambda_2 + \imath \omega x)} \frac{1}{(\lambda_1 + \imath \omega x)} e^{\imath \omega x}$$

$$= \Im \frac{1}{\rho_1 \rho_2} e^{\imath [\omega x + (\phi_1 + \phi_2)]} \tag{9.115}$$

where
$$\rho_k = \sqrt{\lambda_k^2 + \omega^2}, \quad \tan \phi_k = \frac{\omega}{\lambda_k} \tag{9.116}$$

$$\therefore y^\dagger = \frac{1}{\rho_1 \rho_2} \sin \left[\omega x - (\phi_1 + \phi_2) \right]. \tag{9.117}$$

It is easily verified that

$$\rho_1 \rho_2 = \sqrt{(c^2 - \omega^2)^2 + 4b^2 \omega^2}, \tag{9.118}$$

and

$$\phi_1 + \phi_2 = \frac{2b\omega}{c^2 - \omega^2}. \tag{9.119}$$

We note finally that we can use *Method-1* by resolving the right hand side of Eq. (9.114) into partial fractions:

$$y^\dagger = \Im \frac{1}{(D+\lambda_1)(D+\lambda_2)} e^{\iota\omega x} \tag{9.120}$$

$$= \Im \frac{1}{\lambda_2 - \lambda_1} \left[\frac{1}{D+\lambda_1} - \frac{1}{D+\lambda_2} \right] e^{\iota\omega x} \tag{9.121}$$

$$= \Im \frac{1}{\lambda_2 - \lambda_1} \left[\frac{e^{\iota\omega x}}{D+\lambda_1} - \frac{e^{\iota\omega x}}{D+\lambda_2} \right] \tag{9.122}$$

$$= \Im \frac{1}{\lambda_2 - \lambda_1} \left[\frac{e^{\iota\omega x}}{\lambda_1 + \iota\omega} - \frac{e^{\iota\omega x}}{\lambda_2 + \iota\omega} \right] \tag{9.123}$$

$$= \Im \frac{1}{\lambda_2 - \lambda_1} \left[\frac{1}{\lambda_1 + \iota\omega} - \frac{1}{\lambda_2 + \iota\omega} \right] e^{\iota\omega x} \tag{9.124}$$

$$= \Im \frac{1}{(\lambda_1 + \iota\omega)(\lambda_2 + \iota\omega)} e^{\iota\omega x} \tag{9.125}$$

$$= \frac{1}{\rho_1 \rho_2} \sin\left[\omega x - (\phi_1 + \phi_2) \right]. \tag{9.126}$$

Protagonists of the symbolic method

Oliver Heaviside spent most of his life struggling not only with those who could understand neither his mathematics nor the results he obtained but also with most professional mathematicians, who understood the results, and accepted most of them, but disapproved of the devious ways by which he could make differential equations deliver dazzling results that they were not able to extract themselves. Eventually fame without fortune came to him, but by that time his brilliant mind had begun to lose its sharpness and he had become a total recluse. Posterity has tried to recompense him for the injustice, and he is often credited with results that had already been found by his predecessors; some seem to think that the operational method originated with the eccentric Heaviside. For a thorough and scholarly assessment of Heaviside's work, the reader is referred to the text of a lecture delivered by Lionel Cooper [128]. I need only add one publication [129] which does not appear among the works cited by Cooper. Its author, Rehuel Lobatto (1797–1866), used the symbol ∂ instead of D, and spelled out many important relation, including Eq. (9.39); it is worth noting that, in an earlier publication

[130], the same author used the symbol ∂ for denoting an ordinary differential.

In the English speaking world, the symbolic method for solving linear differential equations with constant coefficients was first presented by D. F. Gregory [131], who died aged only 37. His 1839 article can be read by any undergraduate today, and it is both surprising and regrettable that more effort has not been spent on bringing the contents of this paper to the attention of undergraduates.

Part II: Historical

Chapter 10

British Mathematics After Newton

M ANY of those who are aware of the astounding contributions made by Newton may find it hard to believe that British mathematics went into a near-total eclipse soon after Newton's departure from Cambridge, and remained in that pitiable state for almost a century.

Charles Babbage, an undergraduate at Cambridge, stated in the preface to the *Memoirs of the Analytical Society* [132, p. iv]: "But, as if the soil of this country [Britain] were unfavourable to its cultivation [it=calculus], it soon drooped and almost faded into neglect, and we have now to re-import the exotic, with nearly a century of foreign improvement, and to render it once more indigenous among us." It was Babbage who undertook the task of re-importing calculus to Britain and grafting it into the curriculum of Cambridge University, and he found two like-minded and like-brained fellow students willing to join forces with him.

Why did Newton's influence die out so quickly, and why did the eclipse last so long?

In this chapter, I reproduce the views of two British mathematicians who were able to look objectively at the decline of mathematical research in their country: one excerpt is from an article written in 1805 by Rev. John Toplis [133]; the other, a considerably longer passage, is from an anonymous review of Laplace's *Mécanique Céleste* in the *Edinburgh Review* [134], and later in the collected works of John Playfair [135, vol. 4, pp. 261–330].

John Toplis laments

It is a subject of wonder and regret to many, that this island, after having astonished Europe by the most glorious display of talents in mathematics and the sciences dependent upon them, should suddenly suffer its ardour to cool, and almost entirely to neglect those studies in which it infinitely excelled all other nations. After having made the most wonderful and unhoped-for discoveries, and pointed out the road to more; suddenly to desist, and leave these to be cultivated, and the road to more to be explored, by other nations, is very remarkable. It seems as strange as the conduct of a conqueror would be, was he to conquer all the countries around him, and then tamely to suffer his own and the subjugated ones to be possessed, governed, and cultivated, by those whom he had conquered.

It is a very great disgrace for a nation like this, which can proudly boast of a superiority over all others in arts, arms, and commerce, to suffer the sublimest sciences, which once were its greatest pride and glory, to be neglected. Surely a much more solid fame accrues to a people from their superiority in talents than in arms. Athens is as celebrated for its learning as its commerce or its victories. It cannot be owing to any want of importance in the sciences themselves that they are neglected; the discoveries made in them are of the most astonishing nature, and such as seemed absolutely beyond the reach of human intellect. By the marvellous assistance of the mathematics from the simple law of gravity are deduced the orbits of the planets and satellites, their distances, the times of their revolutions, their densities, quantities of matter, and many other remarkable properties too well known to be enumerated. Were it not for them, mechanics, optics, hydrostatics, geography, and other branches of natural philosophy, would hardly have been known as sciences.

John Playfair's diagnosis

Another reflection, of a very different kind from the preceding, must present itself, when we consider the historical details concerning the progress of physical astronomy that have occurred in the foregoing pages. In the list of the mathematicians and philosophers, to whom that science, for the last sixty or seventy years, has been indebted for its improvements, hardly a name from Great Britain falls to be mentioned. What is the reason of this? and how comes it, when such objects were in view, and when so much reputation was to be gained, that the country of Bacon and Newton looked silently on, without taking any share in so noble a contest? In the short view given above, we have hardly mentioned any but the five principal performers [Clairaut, Euler, D'Alembert, Lagrange, Laplace]; but we might have quoted several others, Fontaine, Lambert, Frisi, Condorcet, Bailly, &c. who contributed their share to bring about the conclusion of the piece. In the list, even so extended, there is no British name. It is true, indeed, that before the period to which we now refer, Maclaurin had pointed out an improvement in the method of treating central forces, that has been of great use in all the investigations that have a reference to that subject. This was the resolution of the forces into others parallel to two or to three axes given in position and at right angles to one another. In the controversy that arose about the motion of the apsides in consequence of Clairaut's deducing from theory only half the quantity that observation had established, as already stated, Simpson and Walmsley took a part; and their essays are allowed to have great merit. The late Dr Matthew Stewart also treated the same subject with singular skill and success, in his Essay on the Sun's Distance. The same excellent geometer, in his Physical Tracts, has laid down several propositions that had for their object the determination of the Moon's irregularities. His demonstrations, however, are all geometrical; and leave us to regret, that a mathematician of so much originality preferred the elegant methods of the ancient geometry to the more powerful analysis of modem algebra. Beside these, we recollect no other names of our countrymen distinguished in the researches of physical astronomy during this period; and of these none made any attempt toward the solution of the great problems that then occupied the philosophers

and mathematicians of the Continent. This is the more remarkable, that the interests of navigation were deeply involved in the
question of the lunar theory; so that no motive, which a regard to
reputation or to interest could create, was wanting to engage the
mathematicians of England in the inquiry. Nothing, therefore,
certainly prevented them from engaging in it, but consciousness
that, in the knowledge of the higher geometry, they were not on
a footing with their brethren on the Continent. This is the conclusion which unavoidably forces itself upon us, and which will be
but too well confirmed by looking back to the particulars which
we stated in the beginning of this review, as either essential or
highly conducive to the improvements in physical astronomy.

The calculus of the sines was not known in England till within
these few years. Of the method of partial differences, no mention,
we believe, is yet to be found in any English author, much less
the application of it to any investigation. The general methods of
integrating differential or fluxionary equations, the criterion of
integrability, the properties of homogeneous equations, &c. were
all of them unknown; and it could hardly be said, that, in the
more difficult parts of the doctrine of Fluxions, any improvement
had been made beyond those of the inventor. At the moment
when we now write, the treatises of Maclaurin and Simpson, are
the best which we have on the fluxionary calculus, though such a
vast multitude of improvements have been made by the foreign
mathematicians, since the time of their first publication. These
are facts, which it is impossible to disguise; and they are of such
extent, that a man may be perfectly acquainted with every thing
on mathematical learning that has been written in this country,
and may yet find himself stopped at the first page of the works
or Euler or D'Alembert. He will be stopped, not from the difference of the fluxionary notation, (a difficulty easily overcome,)
nor from the obscurity of these authors, who are both very clear
writers, especially the first of them, but from want of knowing
the principles and the methods which they take for granted as
known to every mathematical reader. If we come to works of still
greater difficulty, such as the Mécanique Céleste, we will venture
to say, that the number of those in this island, who can read that
work with any tolerable facility, is small indeed. If we reckon
two or three in London, and the military schools in its vicinity,

the same number at each of the two English Universities, and perhaps four in Scotland, we shall not hardly exceed a dozen; and yet we are fully persuaded that our reckoning is beyond the truth.

If any further proof of our inattention to the higher mathematics, and our unconcern about the discoveries of our neighbours were required, we would find it in the commentary on the works of Newton, that so lately appeared. Though that commentary was the work of a man of talents, and one who, in this country, was accounted a geometer, it contains no information about the recent discoveries to which the Newtonian system has given rise; not a word of the problem of the Three Bodies, of the disturbances of the planetary motions, or of the great contrivance by which these disturbances are rendered periodical, and the regularity of the system preserved. The same silence is observed as to all the improvements in the integral calculus, which it was the duty of a commentator on Newton to have traced to their origin, and to have connected with the discoveries of his master. If Dr Horsley has not done so, it could only be because he was unacquainted with these improvements, and had never studied the methods by which they have been investigated, or the language in which they are explained.

At the same time that we state these facts as incontrovertible proofs of the inferiority of the English mathematicians to those of the Continent, in the higher department; it is but fair to acknowledge, that a certain degree of mathematical science, and indeed no inconsiderable degree, is perhaps more widely diffused in England, than in any other country of the world. The Ladies' Diary, with several other periodical and popular publications of the same kind, are the best proofs of this assertion. In these, many curious problems, not of the highest order indeed, but still having a considerable degree of difficulty, and far beyond the mere elements of science, are often to be met with; and the great number of ingenious men who take a share in proposing and answering these questions, whom one has never heard of any where else, is not a little surprising. Nothing of the same kind, we believe, is to be found in any other country. The Ladies' Diary has now been continued for more than a century; the poetry, enigmas, &c. which it contains, are in the worst taste possible and the wraps of literature and philosophy are so childish or so old-fashioned, that one is very much at a loss to form a notion of the class of readers to whom they are addressed. The geometrical part, however, has always been conducted in a superior style the problems proposed have tended to awaken curiosity, and the solutions to convey instruction in a much better manner than is always to be found in more splendid publications. If there is a decline, therefore, or a deficiency in mathematical knowledge in this country, it is not to the genius of the people, but to some other cause, that it must be attributed.

An attachment to the synthetical methods of the old geometers, in preference to those that are purely analytical, has often been assigned as the cause of this inferiority of the English mathematicians since the time of Newton. This cause is hinted at by several foreign writers, and we must say that we think it has had no inconsiderable effect. The example of Newton himself may have been hurtful in this respect. That great man, influenced by the prejudices of the times, seems to have thought that algebra and fluxions might be very properly used in the investigation of truth, but that they were to be laid aside when truth was to be communicated, and synthetical demonstrations, if possible, substituted in their room. This was to embarrass scientific method with a clumsy and ponderous apparatus and to reader its progress indirect and slow in an incalculable degree. The controversy that took place, concerning the invention of the fluxionary and the differential calculus, tended to confirm those prejudices, and to alienate the minds of the British from the foreign mathematicians, and the analytical methods which they pursued. That this reached beyond the minds of ordinary men, is clear from the way in which Robins censures Euler and Bernoulli, chiefly for their love of algebra, while he ought to have seen that in the very works which he criticises with so much asperity, things are performed which neither he nor any of his countrymen, at that time, could have ventured to undertake.

We believe, however, that it is chiefly in the public institutions of England that we are to seek for the cause of the deficiency here referred to, and particularly in the two great centres from which knowledge is supposed to radiate over all the rest of the island. In one of these, where the dictates of Aristotle are still listened to

as infallible decrees, and where the infancy of science is mistaken
for its maturity, the mathematical sciences have never flourished;
and the scholar has no means of advancing beyond the mere elements of geometry. In the other seminary, the dominion of prejudice is not equally strong; and the works of Locke and Newton
are the text from which the prelections are read. Mathematical
learning is there the great object of study; but still we must dis-
approve of the method in which this object is pursued. A certain
portion of the works of Newton, or of some other of the writers
who treat of pure or mixed mathematics in the synthetic method,
is prescribed to the pupil, which the candidate for academical
honours must study day and night. He must study it, not to learn
the spirit of geometry, or to acquire the $\delta\upsilon\alpha\mu\iota\varsigma\ \varepsilon\upsilon\rho\eta\tau\iota\kappa\eta$ by which
the theorems were discovered, but to know them as a child does
his catechism, by heart, so as to answer readily to certain inter-
rogations. In all this, the invention finds no exercise; the student
is confined within narrow limits; his curiosity is not roused; the
spirit of discovery is not awakened. Suppose that a young man,
studying mechanics, is compelled to get by heart the whole of
the heavy and verbose demonstrations contained in Keil's introduction, (which we believe is an exercise sometimes prescribed,)
what is likely to be the consequence? The exercise afforded to the
understanding by those demonstrations, may no doubt be im-
proving to the mind; but as soon as they are well understood, the
natural impulse is to go on; to seek for something higher; or to
think of the application of the theorems demonstrated. If this natural expansion of the mind is restrained, if the student is forced
to fall back, and to go again and again over the same ground,
disgust is likely to ensue; the more likely, indeed, the more he
is fitted for a better employment of his talents; and the least evil
that can be produced, is the loss of the time, and the extinction of
the ardour that might have enabled hint to attempt investigation
himself, and to acquire both the power and the taste of discovery.
Confinement to a regular routine, and moving round and round
in the same circle, must, of all things, be the most pernicious to
the inventive faculty. The laws of periodical revolution, and of
returning continually in the same track, may, as we have seen, be
excellently adapted to a planetary system, but are ill calculated to
promote the ends of an academical institution. We would wish to

see, then, some of those secular accelerations, by which improvements go on increasing from one age to another. But this has been rarely the case; and it is melancholy to reflect, how many of the universities of Europe have been the strong holds where prejudice and error made their last stand—the fastnesses from which they were latest of being dislodged. We do not mean to hint that this is true of the university of which we now speak, where the credit of teaching the doctrines of Locke and Newton is sufficient to cover a multitude of sins. Still, however, we must take the liberty to say, that Newton is taught there in the way least conducive to solid mathematical improvement.

Perhaps, too, we might allege, that another public institution, intended for the advancement of science, the Royal Society, has not held out. In the course of the greater part of the last century, sufficient encouragement for mathematical learning. But this would lead to a long disquisition: And we shall put an end to the present digression, with remarking, that though the mathematicians of England have taken no share in the deeper researches of physical astronomy, the observers of that country have discharged their duty better. The observations of Bradley and Maskelyne have been of the utmost importance in this theory; their accuracy, their number, and their uninterrupted series, have rendered them a fund of immense astronomical riches. Taken in conjunction with the observations made at Paris, they have furnished Laplace with the *data* for fixing the numerical values of the constant quantities in his different series; without which, his investigations could have had no practical application. We may add that no man has so materially contributed to render the formulas of the mathematician useful to the art of the navigator, as the present astronomer royal. He has been the main instrument of bringing down this philosophy from the heavens to the earth; of adapting it to the uses of the unlearned and of making the problem of the Three Bodies the surest guide of the mariner in his journey across the ocean.

Who brought calculus back to Britain

Far more surprising than the decline of British mathematics after Newton is the story of the conspiracy which brought calculus

back to Britain. The conspirators were three brilliant undergraduates: Charles Babbage (1791–1871), John Herschel (1792–1871),
and George Peacock. (1791–1858). The story has been recounted
with evident relish by Babbage, the conspirator-in-chief, in his
memoirs, an excerpt from which appears below:

At this period Cambridge was agitated by a fierce controversy.
Societies had been formed for printing and circulating the Bible.
One party proposed to circulate it with notes, in order to make
it intelligible; whilst the other scornfully rejected all explanations
of the word of God as profane attempts to mend that which was
perfect.

The walls of the town were placarded with broadsides, and
posters were sent from house to house. One of the latter form of
advertisement was lying upon my table when Slegg left me. Taking up the paper, and looking through it, I thought it, from its ex-
aggerated tone, a good subject for a parody.

I then drew up the sketch of a society to be instituted for trans-
lating the small work of Lacroix on the Differential and Integral
Lacroix. It [was] proposed that we should have periodical meetings
for the propagation of d's; and consigned to perdition all who supported the heresy of dots. It maintained that the work of Lacroix
was so perfect that any comment was unnecessary.

On Slegg's return from chapel I put the parody into his hands.
My friend enjoyed the joke heartily, and at parting asked my per-
mission to show the parody to a mathematical friend of his, Mr.
Bromhead.

The next day Slegg called on me, and said that he had put the
joke into the hand of his friend, who, after laughing heartily, remarked that it was too good a joke to be lost, and proposed seriously that we should form a society for the cultivation of mathematics.

The next day Bromhead called on me. We talked the subject
over, and agreed to hold a meeting at his lodgings for the purpose
of forming a society for the promotion of analysis.

At that meeting, besides the projectors, there were present Herschel, Peacock, D'Arblay, Ryan, Robinson, Frederick Maule, and
several others. We constituted ourselves "The Analytical Society;"
hired a meeting-room, open daily; held meetings, read papers, and
discussed them. Of course we were much ridiculed by the Dons;
and, not being put down, it was darkly hinted that we were young
infidels, and that no good would come of us.

152

In the meantime we quietly pursued our course, and at last re-
solved to publish a volume of our Transactions. Owing to the ill-
ness of one of the number, and to various other circumstances, the
volume which was published was entirely contributed by Herschel
and myself.

At last our work was printed, and it became necessary to decide
upon a title. Recalling the slight imputation which had been made
upon our faith, I suggested that the most appropriate title would
be—

The Principles of pure D-ism in opposition to the Dot-age of the
University.

I prefer the title as it was (mis)quoted by Newman [136, p. 59]:
"...pure *d*-ism in opposition to the dot-age ...".

The translation, together with an appendix and a set of notes,
was published in 1816, and I will baptize it as the *Cambridge
Lacroix*. The first paragraph of Note B is reproduced below [137,
pp. 569–70]:

> The method which is made use of by our author, in the expo-
> sition and demonstration of the principles of the Differential Cal-
> culus, was first given by D'Alembert, in the Encyclopédie.* [the
> asterisk refers to the footnote: Art. Différentiel] We shall not at
> present stop to discuss its merits, but shall proceed directly to shew
> in what manner this calculus may be established upon principles
> which are entirely independent of infinitesimals or limits. We shall
> afterwards endeavour to explain the reason why the same conclu-
> sions have invariably been deduced from apparently different first
> principles.
>
> It is hardly necessary to inform the reader, that we are indebted
> for the principal part of the contents of this note to the Calcul des
> Fonctions of Lagrange, and the large treatise by our author, on the
> Differential and Integral Calculus.

In other words, the brand of calculus which the gang of three
imported to Britain was soon going to be replaced by one where
the derivative would once again be defined, but with greater attention to rigour, as the limit of a ratio of two increments. Readers who are not familiar with the development of the concept of
derivative are referred to Kline's book [138, pp. 954–6].

The young rebels knew that publishing a book was only the

first step towards Continentalization of British mathematics, and that the next move would be ineffective unless it was carried out with a carrot in one hand and a stick in the other, which meant, in this situation, publishing a compilation of useful formulas as a supplement to the *Cambridge Lacroix* and including, in the final examination paper, questions phrased in *d*-istic jargon. These responsibilities were handled admirably by George Peacock, and those who are interested in details are referred to two obituaries [139, 140]. For reasons that will soon become apparent, it is important to quote a passage from one of these notices [140]:

> In 1817 he was Moderator, and in this capacity he ventured to introduce the new system into the public examinations, his colleagues retaining the old one. This old system made its appearance once more in 1818; in 1819 Mr Peacock was Moderator, with a colleague of his own sentiments (Mr. Gwatkin), and the change was fully accomplished. All the chief actors in producing it have lived to see their work fully done, and *their country in full communication with all the world after more than a century of nearly complete exclusion.* (Emphasis added.)

Airy's *Tracts*

George Biddell Airy (1801–92), who graduated from Trinity College in 1823, became Lucasian professor of mathematics in 1826, and Plumian professor of astronomy and director of the Cambridge observatory in 1828. In 1835 he became the seventh astronomer royal (director of the Royal Greenwich Observatory), a post he held for more than 45 years.

Airy wrote several books for the use of undergraduates. The first book, entitled *Mathematical Tracts on ...*, went through four editions (1826, 1831, 1842 and 1858).

In the preface to the first edition, Airy wrote [56, p. viii]:

> The author has to apologize for the introduction of an uncommon symbol of integration. For the student who confines himself to the use of differential coefficients, it appeared necessary to employ a symbol which should not require the use of a differential: no confusion, it is imagined, will be occasioned to those who do not adopt that system. The definition of a differential coefficient referred to throughout the work is "the limit of the ratio of the corresponding increments of the function and the independent variable."

Airy began the text by explaining his notation for the integral [56, p. 1]:

By the notation $\int_\theta \cos n\theta \cdot \Theta$, $\int_\theta \cos n\theta \cdot \cos \overline{m\theta + D}$,&c. we mean what are usually written

$$\int_\theta \cos n\theta \cdot \Theta \cdot d\theta, \quad \int_\theta \cos n\theta \cdot \cos \overline{m\theta + D} \cdot d\theta, \&c.$$

they are the quantities whose differential coefficients, with respect to θ, are $\cos n\theta \cdot \Theta$, $\cos n\theta \cdot \cos \overline{m\theta + D}$, &c.

When Airy had to deal with what we now call a Riemann sum, he used the same symbol and stated the limits of the integral either in words, or in an array after the integrand, as in the equation displayed below, which is taken from one of his papers [141]:

$$\frac{2}{n^3 a^2} \int_x \frac{Fx}{\sqrt{a^2 - x^2}} \begin{Bmatrix} x = c \\ x = a \end{Bmatrix} = \frac{2F}{n^3 a^2} \sqrt{a^2 - c^2}.$$

Whewell's *Doctrine of Limits*

William Whewell (1794–1866), a real polymath and one of the most influential British scholars of his time, wrote on such a wide variety of subjects that the satirist Sydney Smith could not resist saying "Science was his forte, omniscience his foible" [142]. Whewell also had a knack for coining scientific terms, and it is he who gave us the word *scientist*.

This is how Whewell began his chapter on integration in *The Doctrine of Limits*, which was published in 1838 [143, p. 151]:

1. If y, p be functions of x, such that $\frac{dy}{dx} = p$, y is the Integral of p.

2. This relation is denoted thus;

$$y = \int_x p, \quad \text{or} \quad y = \int p \, dx$$

3. The *Integral Calculus* is a system of rules and processes, the object of which is to find the integrals of any functions.

4. The fundamental rules of integration are derived from the rules of differentiation.

$$\text{If} \qquad y = \frac{x^{m+1}}{m+1}, \quad \text{we have} \quad \frac{dy}{dx} = x^m; \qquad (10.1)$$

$$\text{therefore} \qquad \int_x x^m = \frac{x^{m+1}}{m+1}. \qquad (10.2)$$

5. If $y = \dfrac{1}{m \cdot n + 1 b}(a + bx^m)^{n+1}; \quad \dfrac{dy}{dx} = (a + bx^m)^n x^{m-1};$

$$\text{therefore} \int_x (a + bx^m)^n x^{m-1} = \frac{1}{m \cdot n + 1 b}(a + bx^m)^{n+1}$$

6. If $y = -\cos x, \quad \dfrac{dy}{dx} = \sin x; \quad \text{therefore} \int_x \sin x = -\cos x.$

15. If p be any function of x, it may be proved in like manner that

$$\int_x p^m \frac{dp}{dx} = \frac{p^{m+1}}{m+1}.$$

$$\int_x (a + bp^m)^n p^{m-1} \frac{dp}{dx} = \frac{1}{m \cdot n + 1 b}(a + bp^m)^{n+1}.$$

$$\int_x \sin p \frac{dp}{dx} = -\cos p.$$

I add for the sake of forestalling some misunderstanding that items 7–14 in Whewell's list have been dropped and that item 15 has more entries after the last one shown above.

Augustus De Morgan

Augustus De Morgan (1806–71) was a mathematician, a logician, an actuary, a historian, a biographer, a bibliophile, and a peerless expositor of mathematics, both as a lecturer (in the opinion of those who had the privilege of listening to him) and as the author of several books, most of which I have read, including *The Differential and Integral Calculus* (abbreviated hereafter as *D&I*), which remains, at least in my opinion, one of the best books on the topic written in English, but I am sure that I would not have been so generous in my praise if I had found a copy in my under-graduate days. An Obituary writer remarked [144]: "As a writer of mathematical text-books he took the highest rank, his books being more suitable, however, for teachers than for pupils." If one reads the preface to the second edition of *D&I*, one imme-diately discovers that the students De Morgan addressed were not those who just wanted to get respectable grades in the final examination. The preface begins thus:

> The work now before the reader is the most extensive which our language contains on the subject, being (exclusive of the Elementary Illustrations, at the end) more than double in matter of the Cambridge translation of Lacroix, and full half as much as the great work of the same author in three volumes quarto. I state this because *students are sometimes apt to be discouraged by the apparent slowness of their progress, which they measure by the pages read, without any other consideration.* (Emphasis added.) This extent of matter is not due to fullness of explanation or abundance of examples, but to a variety of subjects exceeding that which is usually introduced into elementary writings. There are many works which enter more largely into the simpler parts, and elucidate them by more copious instances, both of which my specific object has prevented me from doing. That object has been to contain, within the prescribed limits, the whole of the student's course, from the confines of elementary algebra and trigonometry, to the entrance of the highest works on mathematical physics. A learner who has a good knowledge of the subjects just named, and who can master the present treatise, taking up elementary works on conic sections, application of algebra to geometry, and the theory of equations, as he wants them, will, I am perfectly sure, find himself able to conquer the difficulties of anything he may meet with; and need not close any book of Laplace, Lagrange, Legendre, Poisson, Fourier, Cauchy, Gauss, Abel, Hindenburgh and his followers, or of any one of our English mathematicians, under the idea that it is too hard for him. It may be admitted to be desirable that some one writer should endeavour to attain such a result as that of placing before the student all that is requisite to put him in communication with the highest investigators; and it will readily be seen, that unless a very large work indeed were written, no such result could be obtained without condensation, particularly in the higher parts. If much difficulty should be experienced in the elementary

chapters, I know of no work which I can so confidently recommend to be used with the present one, as that of M. Duhamel, cited in the note to page 681.

De Morgan also wrote *Elementary Illustrations of the Differential and Integral Calculus*, a primer which appeared in two instalments (15 September and 15 November 1832, numbers 135 and 140) in the Library of Useful Knowledge. This booklet, abbreviated hereafter as *Elementary Illustrations*, is usually bound together with *D&I*, which too was issued serially at first, and as a single volume in 1836 [89]; the second edition was published six years later [51]. In the first edition, the prefatory part is entitled *Advertisement*, and *Elementary Illustrations* appears before *D&I*; in the second edition, p. ii has a note "To The Binder", with the instruction: "The Elementary Illustrations (64 pages) are to be bound at the end of the work, after the Table of Errata; and the Title, which was printed in the first number of the work, is to be cancelled, but the Advertisement appended to it, to remain in its place." Not all copies of the second edition contain *Elementary Illustrations*, but, since *Elementary Illustrations* and *D&I* are paginated independently, it does not matter where the former is located in a bound volume that contains both works.

In the *Advertisement* the author emphasized that his aim was to make the theory of limits "the sole foundation of the science [of calculus], without any aid whatsoever from the theory of series, or algebraical expansions", and that no existing book in English had attempted to do this. He also defended his decision to treat the two facets of calculus (differentiation and integration) at the same time.

Let us begin by reproducing two paragraphs from p. 34 of *D&I* in either edition:

> The notation of Newton, which prevailed in England till after the commencement of the present century, has been discarded by all writers in the universities, and by most out of them. There are those who object to the change, and who consider the fluxional notation as at least equal, if not superior, to that of Leibnitz. Without discussing this point, we are inclined to consider the universality of the notation of Leibnitz throughout the whole of the civilized world, and the fact of most of the discoveries made since the time of Newton, both

in pure mathematics and physics, being expressed by means of it, as itself a sufficient reason for adopting it. But we shall in the proper place give both notations, and explain the method of converting one into the other. [We will see later (Chapter 13) that De Morgan is not equally tolerant of Airy's notation.]

We shall also endeavour to teach the Integral Calculus at the same time as the Differential. The separation of the two which takes place in most works, though convenient in some respects, and those not unimportant, yet deprives the student of the means of learning, at the same time, subjects between which the analogy is as strong as between addition and subtraction.

As mentioned above, and also discussed later, De Morgan did not take kindly to Airy's symbol $\int_x$ and its counterpart d_x. It is reasonable to suppose that De Morgan and the others who opposed the so-called Cambridge notation were trepidatious that, just when British mathematicians had returned to the mathematical fold, they would become pariahs once more if they succumbed to a bout of separatism. De Morgan explains the vital importance of notational ecumenism in *Elementary Illustrations* (p. 60):

> As in other cases, the expressions of Leibnitz are the most convenient and the shortest, for all who can immediately put a rational construction upon them; this, and the fact that, good or bad, they have been, and are, used in the works of Lagrange, Laplace, Euler, and many others, which the student who really desires to know the present state of physical science, cannot dispense with, must be our excuse for continually bringing before him modes of speech, which, taken quite literally, are absurd.

Chapter 11

Attempts to Algebrify Calculus

T HE term *algebrification of calculus* should not be confused with the process *algebrization of operators* which was discussed in Chapter 9. When applied to the foundations of calculus, algebrification means an attempt to rid calculus of the notion of infinitesimals and concepts of motion; though this project is now considered to have been misguided and therefore (pre)destined to fail, its study is not devoid of considerable pedagogical value.

John Landen (1719–1790) and *Residual Analysis*

The first attempt to algebrify calculus seems to have come from John Landen [145, 146], who earned his living as a surveyor, and used his leisure hours for cultivating his mind through mathematical investigations. For various reasons that need not be discussed here, his work did not attract any devotees in England, but won praise from several continental mathematicians.

In presenting the idea to the public, Landen wrote [147]:

> Yet, notwithstanding the method of fluxions is so greatly applauded, I am induced to think, it is not the most natural method of resolving many problems to which it is usually applied. The operations therein being chiefly performed with algebraic quantities, it

159

is, in fact, a branch of the algebraic art, or an improvement thereof, made by the help of some peculiar principles borrowed from the doctrine of motion: which principles, I must confess, to me seem not so properly applicable to algebra as those on which that art was, before, very naturally founded. We may indeed very naturally conceive a line to be generated by motion; but there are quantities of various kinds, which we cannot conceive to be so generated. It is only in a figurative sense, that an algebraic quantity can be said to increase or decrease with some velocity or degree of swiftness; and, by the fluxion of a quantity of that kind, we mean, I presume, to have a clear idea of its meaning, understand the velocity of a point supposed to describe a line denoting such quantity. Fluxions therefore are not immediately applicable to algebraic quantities; but in fluxionary computations by such quantities, we, to proceed with perspicuity, we must have recourse to the supposition of lines being put to denote those quantities, and the generation of those lines by motion. It therefore, to me, seems more proper, in the investigation of propositions by algebra, to proceed upon the *anciently-received* principles of that art, than to introduce therein, without any necessity, the new fluxionary principles, derived from a consideration of motion; and the rather, as the introduction of those new principles is not attended with any peculiar advantage.—That the borrowing principles from the doctrine of motion, with a view to improve the analytic art, was done, not only without any necessity, but even without any peculiar advantage, will appear by showing, that whatever can be done by the method of computation, which is founded on those borrowed principles, may be done as well, by another method founded entirely on the *anciently-received* principles of algebra: And that I shall endeavour to shew, as soon as I have leisure, in the treatise I lately proposed to publish by subscription.[1]—In the mean time, this essay is intended, to give the inquisitive reader some notion of the new method of computation, which is the subject of that treatise.—Which method I call the *Residual Analysis* because, in all the enquiries wherein it is made use of, the conclusions are obtained by means of residual quantities.

The term *residual quantity*, a part of algebraic vocabulary in

---

[1]Landen is announcing here a slightly larger work that appeared six years later [148], and even promised a sequel that did not appear.

Landen's time, means a difference of two quantities; thus $a - b$ and $a - \sqrt{b}$ are both residual quantities. Some impression of how Landen obtained derivatives may be gained by considering the following expansion

$$\frac{x^m - a^m}{x - a} = x^{m-1} + A_2 x^{m-2} + A_3 x^{m-3} + \cdots a^{m-1},$$

where m is a positive integer, and the A_k are coefficients whose values are to be determined. The validity of the expansion may be judged by noting that when $x = 0$, the right-hand side becomes a^{m-1}, and when $a = 0$, the right-hand side becomes x^{m-1}, as it ought. To determine the coefficients, we multiply both sides with $(x - a)$, and compare the coefficients of various powers of x; this leads to the conclusion

$$A_2 - a = A_3 - A_2 a = A_4 - A_3 a \cdots = 0, \qquad (11.1)$$
$$\therefore A_k = a^{k-1}, \qquad (2 \le k \le m - 2). \qquad (11.2)$$

Hence

$$\frac{x^m - a^m}{x - a} = x^{m-1} + a x^{m-2} + a^2 x^{m-3} + \cdots a^{m-1},$$

If we put $x = a$, the left-hand side becomes a vanishing fraction, whereas the right-hand side becomes $m x^{m-1}$. According to Landen, the value of the right-hand side for $x = a$ is the value to be assigned to the vanishing fraction, and implies (or justifies) the relation $dx^m = (m - 1)x^{m-1} dx$.

For a fair assessment of residual analysis, I quote Leslie [149, p. 601]: "Landen, one of those men so frequent in England [and elsewhere?] whose talents surmount their narrow education, produced, in 1758, a new form of the Fluxionary Calculus, under the title of Residual Analysis, which, though framed with little elegance, may be deemed, on the whole, an improvement on the method of ultimate ratios. To confer more consequence on his innovation, he contrived likewise a set of symbols, and applied his algorithm to the solution of different problems. But it never obtained any currency, and soon fell into oblivion."

Joseph-Louis Lagrange (1736–1813) and *Theorie des Fonctions Analytiques* (1797)

Lagrange's aim can be grasped by examining the complete title of his 1797 book, which may be roughly rendered as follows: *Theory of analytic functions, containing the principles of differential calculus, free from all consideration of infinitesimals, vanishing quantities, limits and fluxions, and reduced to algebraic analysis of finite quantities.* His strategy is summarized below.

Let $u = f(x)$ express the relation between the function and its independent variable x. Let x increase and become $x + h$, then the value of u will most probably be altered. Let the new value be represented by U, with $U = f(x+h)$ and $u = f(x)$, the difference comes out to be

$$U - u = f(x+h) - f(x). \qquad (11.3)$$

We follow Lagrange and begin by writing

$$f(x+h) = f(x) + h\, P_0(x,h), \qquad (11.4)$$

where P_0 is a new function of x and h; he went on to separate what is independent of h, and therefore does not vanish when h becomes zero, and a part that survives when $h = 0$. By following a similar reasoning as above, we set

$$P_0 = p_1(x) + hP_1(x,h), \qquad (11.5)$$

where p_1 is a new function of x, and P_1 a new function of x and h that does not become infinite when $h = 0$.

By proceeding recursively and setting

$$P_1 = p_2(x) + hP_2(x,h), \quad P_2 = p_3(x) + hP_3(x,h),\ldots \qquad (11.6)$$

we arrive at the result

$$f(x+h) = f(x) + p_1 h + p_2 h^2 + p_3 h^2 + \ldots \qquad (11.7)$$

The recursion scheme consists of the following steps

$$hP_0 = f(x+h) - f(x) \qquad (11.8a)$$

$$hP_k = P_{k-1} - p_k \qquad (k \geq 1) \qquad (11.8\text{b})$$

$$p_k = [P_{k-1}]_{h=0} \qquad (11.8\text{c})$$

In *Théorie des fonctions analytiques* [150] Lagrange gave some examples of this recursive procedure, the first being $f(x) = 1/x$ (p. 9). In this case, we find

$$hP_0 = \frac{1}{x+h} - \frac{1}{x} = -\frac{h}{x(x+h)}; \qquad P_0 = -\frac{1}{x(x+h)}; \qquad p_1 = -\frac{1}{x^2},$$

$$hP_1 = -\frac{1}{x(x+h)} + \frac{1}{x^2} = \frac{h}{x^2(x+h)}; \qquad P_1 = \frac{1}{x^2(x+h)}; \qquad p_2 = \frac{1}{x^3},$$

$$hP_2 = \frac{1}{x^2(x+h)} - \frac{1}{x^3} = -\frac{h}{x^3(x+h)}; \qquad P_2 = -\frac{1}{x^3(x+h)}; \qquad p_3 = -\frac{1}{x^4}.$$

For the next case considered by Lagrange, namely $f(x) = \sqrt{x}$, the first three stages of the calculation proceed as follows:

$$hP_0 = \sqrt{x+h} - \sqrt{x} = -\frac{h}{\sqrt{x+h} + \sqrt{x}}$$

$$P_0 = \frac{1}{\sqrt{x+h} + \sqrt{x}}$$

$$p_1 = \frac{1}{2\sqrt{x}};$$

$$hP_1 = P_0 - p_1 = \frac{1}{\sqrt{x+h} + \sqrt{x}} - \frac{1}{2\sqrt{x}} = \frac{\sqrt{x} - \sqrt{x+h}}{2\sqrt{x}\left[\sqrt{x+h} + \sqrt{x}\right]}$$

$$= -\frac{1}{2\sqrt{x}\left[\sqrt{x+h} + \sqrt{x}\right]^2}h$$

$$P_1 = -\frac{1}{2\sqrt{x}\left[\sqrt{x+h} + \sqrt{x}\right]^2}$$

$$p_2 = -\frac{1}{8x\sqrt{x}};$$

$$hP_2 = P_1 - p_2 = \frac{h}{2\sqrt{x}} \cdot \frac{\sqrt{x+h} + 3\sqrt{x}}{4x\left[\sqrt{x+h} + \sqrt{x}\right]^3}$$

$$P_2 = \frac{1}{2\sqrt{x}} \cdot \frac{\sqrt{x+h} + 3\sqrt{x}}{4x\left[\sqrt{x+h} + \sqrt{x}\right]^3}$$

$$p_3 = \frac{1}{16x^2\sqrt{x}};$$

Let us note that a function $u = f(x)$ which is such that $U = f(x + h)$ can be expanded in the manner shown in Eq. (11.7) is said to be *differentiable* (see Appendix A for a discussion of the concept of differentiability). We see that if

$$U - u = p_1 + p_2 h^2 + \cdots , \quad \text{then} \qquad (11.9)$$

$$\frac{U - u}{h} = p_1 + p_2 h + \cdots \qquad (11.10)$$

$$\therefore D_x u \equiv \lim_{h \to 0} \frac{U - u}{h} = p_1 \qquad (11.11)$$

Criticism of Lagrangian view of calculus

Although no one now advocates the de-extinction of the original version of Langrangian calculus, it will be instructive to read some critical comments, and I have chosen the views of Augustus De Morgan and Felix Klein in this section.

> **De Morgan** [51, pp. iv–v]: The method of Lagrange, founded on a very defective demonstration of the possibility of expanding $\phi(x + h)$ in whole powers of h, had taken deep root in elementary works; it was the sacrifice of the clear and indubitable principle of limits to a phantom, the idea that an algebra without limits was purer than one in which that notion was introduced. But, independently of the idea of limits being absolutely necessary even to the proper conception of a convergent series, it must have been obvious enough to Lagrange himself, that all application of the science to concrete magnitude, even in his own system, required the theory of limits. Some time after the publication of the first numbers of this work, four[2] different treatises appeared in the French language, all of which rejected the doctrine of series, and adopted that of limits. I have therefore no occasion to argue further against the former method, which has been thus abandoned in the country which saw its birth, and will certainly lose ground in England, when it is no longer maintained by a supply from abroad of elementary treatises written on its principles.

---

[2]De Morgan inserts a footnote here, citing four recently published book [151–154], and adds that he has not yet seen the last.

Klein [55, p. 220]: This reaction against differentials accounts also for the attempt by Lagrange, already mentioned, in his Theorie des Fonctions Analytiques, published in 1797, to eliminate from the theory not only infinitely small quantities, but also every passage to the limit. He confined himself, namely, to those functions which are defined by power series

$$f(x) = a_0 + a_1x + a_2x^2 + a_3x^3 + \cdots.$$

and he defines formally the "derived function $f'(x)$" (he avoids characteristically the expression differential quotient and the sign dy/dx) by means of a new power series

$$f'(x) = a_1 + 2a_2x + 3a_3x^2 + \cdots.$$

Consequently he talks of *derivative calculus* instead of *differential calculus*. This presentation, of course, could not be permanently satisfactory. In the first place, the concept of function used here is, as we have shown, much too limited. More than that, however, such thoroughly formal definitions make a deeper comprehension of the nature of the differential coefficient impossible, and take no account of what we called the *psychological moment*—they leave entirely unexplained just why one should be interested in a series obtained in such a peculiar way. Finally, one can get along without giving any thought to a limit process only by disregarding entirely the convergence of these series and the question within what limits of error they can be replaced by finite sums. As soon as one begins a consideration of these problems, which is essential, of course, for any actual use of the series, it is necessary to have recourse precisely to that notion of limit, the avoidance of which was the purpose of inventing the system.

In Chapter 12 we will take a closer look at the Lagrangian formulation of calculus to see how derivatives are defined in this approach, and how it may be extended to functions of more than one independent variable.

Chapter 12

Earnshaw's Pamphlet

Earnshaw's theorem is much better known than its eponym, the Reverend Samuel Earnshaw (1805–88), and his pamphlet (the topic of this chapter) is all but forgotten. The paper in which he proved the theorem was presented in 1839 but published in 1842 [155]; the first equation in the article is reproduced below:

$$F = d_f V, G = d_g V, H = d_h V.$$

Here F, G, and H are the forces (directed along the coordinate axes) exerted on a body (called the attracted body) by another body (called the attracting body); f, g and h are the coordinates of the attracted body; the symbols d_f, d_g and d_h are left undefined, but an experienced reader would be able to infer their meanings and confirm her/his inference by reading the rest of this landmark paper. The symbol d_x was introduced by Thomas Jarrett (see below), and Earnshaw extolled its virtues in a pamphlet entitled *On the Notation of the Differential Calculus* [156].

Most of this chapter is a paraphrase of the above named pamphlet; the reasons for making the changes will be explained in due course.

A profusion of symbols for partial derivatives and differentials

A neophyte who has read only modern textbooks might not realize the state of confusion created by the plethora of symbols used by previous authors for denoting partial differentiation. For their benefit, I recall three examples from the period 1830–60.

J. R. Young's *Elements of Differential Calculus*

There is a British edition [157] of this book as well as an American edition [158], both published in 1833 and freely available.

§36 ends with the following pair of equations:

$$\frac{du}{dx} = \frac{du}{dx} + \frac{du}{dq} \cdot \frac{dq}{dx} \tag{12.1}$$

$$\frac{du}{dx} = \frac{du}{dx} + \frac{du}{dq} \cdot \frac{dq}{dx} + \frac{du}{dr} \cdot \frac{dr}{dx} \tag{12.2}$$

in the first (second) of which u is a function of x and q (x, q and r), q and r being themselves functions of x. The text of the following article is reproduced below:

> We must not confound here the $\dfrac{du}{dx}$ on the left, with that on the right, in these equations; for the former denotes the *total* differential coefficient, of which the latter forms but a part, and is therefore called a *partial* differential coefficient. It is to be regretted, however, that analysts are not agreed as to the best means of distinguishing total, from partial, differential coefficients; and accordingly, in most works on the calculus, the same symbol is applied indiscriminately to both; a circumstance likely to prove a frequent source of perplexity to the learner; and to avoid which we shall, throughout this volume, always distinguish the total differential coefficient by enclosing it in braces, so that the two equations above will be written thus:

$$\left\{\frac{du}{dx}\right\} = \frac{du}{dx} + \frac{du}{dq} \cdot \frac{dq}{dx} \tag{12.3}$$

$$\left\{\frac{du}{dx}\right\} = \frac{du}{dx} + \frac{du}{dq} \cdot \frac{dq}{dx} + \frac{du}{dr} \cdot \frac{dr}{dx} \tag{12.4}$$

W. C. Ottley's *Treatise on the Differential Calculus* **(1838)**

This is what Ottley has to say [159, p. 87]:

> The quantities . . . are called the *partial differential coefficients* of u with respect to x, y, z respectively, and they are commonly represented by the symbols

$$\left\{\frac{du}{dx}\right\}, \left\{\frac{du}{dy}\right\}, \left\{\frac{du}{dz}\right\}.$$

And on p. 91 one reads the following:

$$\therefore \frac{du}{dx} = \left\{\frac{du}{dx}\right\} + \left\{\frac{du}{dy}\right\}\frac{dy}{dx}$$

where the differential coefficients in the brackets are those that would have been obtained by differentiating it separately, on the suppositions y constant and x constant, respectively; and $\frac{dy}{dx}$ is the general differential coefficient of y as a function of x.

Isaac Todhunter

Todhunter's *Treatise on the Differential Calculus* went through five editions (1852, 1855, 1860, 1864, 1871), the last of which was reprinted several times (1873, 1875, 1878, 1881, 1885, 1890). The material quoted below may be found in §§169–170 in any edition available to the reader.

§169 ends with the equation (which is not numbered by the author)

$$\frac{du}{dx} = \frac{du}{dy}\frac{dy}{dx} + \frac{du}{dz}\frac{dz}{dx}. \tag{12.5}$$

The following article is quoted below:

170. In this result $\frac{du}{dx}$ denotes, as stated, "the differential coefficient of u, taken with respect to y, supposing y alone to vary." It is not impossible that the reader may be inclined to say, "But y and z being both functions of x, if y varies, z *must* vary too, how then *can* I make the supposition that y alone varies?" His own further reflexion will probably remove the difficulty, if such it be. Should he however be unable to satisfy himself, it may be suggested to him that we do not make the supposition that y alone varies as a *final* supposition. We allow for the variation of both y and z, but it is convenient for our purpose to consider these variations one at a time.

It is usual to write $\left(\frac{du}{dz}\right)$ and $\left(\frac{du}{dy}\right)$, instead of $\frac{du}{dx}$ and $\frac{du}{dz}$, the brackets serving to remind us of the suppositions to be made in finding the values of these differential coefficients. Hence the above equation should be written

$$\frac{du}{dx} = \left(\frac{du}{dy}\right)\frac{dy}{dx} + \left(\frac{du}{dz}\right)\frac{dz}{dx}. \tag{12.6}$$

> Of course *the brackets may be omitted, and indeed frequently are omitted,* provided we can feel certain of remembering the conditions which they are designed to express. The beginner will do well to use them, although as he advances in the subject he may be able to dispense with them. (My italics.)

Earnshaw's motive for writing the pamphlet

Most of Earnshaw's contemporaries did not use braces or parentheses for distinguishing the ordinary differential coefficient from its partial counterpart. Not satisfied with a notation which could easily confuse a beginner, Earnshaw thought that the introduction of a new symbol for the differential coefficient would preempt the confusion, and establish the superiority of the calculus of differential coefficients over the calculus of differentials by exposing the weakness of the Leibnizian calculus, where a differential coefficient is defined as a quotient of two differentials. He adopted the symbol d_x, for differentiation with respect to x, which had been proposed in 1827 by the Rev. Thomas Jarrett [160, 161], and wrote $\dfrac{du}{dx}$ only when he dealt with a ratio of two differentials.

In the next section an edited version of Earnshaw's tract is presented. It has been necessary to make some changes in the notation, but no purpose will be served by listing all of them here; suffice it to say, the symbol d_x has been replaced by D_x (for ordinary differentiation) or $\mathscr{D}_x$ (for partial differentiation).

Edited version of Earnshaw's pamphlet

§1. In a text dealing with differential calculus, those quantities which throughout any operation are supposed to retain the same value are called *constants*; but those which are supposed to have, or to be capable of having, different successive values are called *variables*.

Thus the equation of a circle, the co-ordinates of whose centre are α, β, and whose radius is γ, being

$$(x - \alpha)^2 + (y - \beta)^2 = \gamma^2; \qquad (12.7)$$

so long as we are speaking of points in the same circle, x and y may be supposed variable, since they belong equally to any point in the circle, and α, β, γ constant; but if we are speaking of different circles all passing through the same point, (viz. that whose coordinates are x, y), then x and y are constants, and α, β, γ variable; and, finally, if we are speaking of different circles passing through different points (as for instance, through successive points in a given curve), then we must consider x, y, α, β, and γ as variable. It is the nature of a question, therefore, that determines which are the constant and which the variable quantities.

§2. The variableness of one quantity (say y) may wholly depend on that of another (say x), and in this case y is said to be a function of x only: but if the variableness of y depend on that of v, as well as on that of x, while the variableness of v does not depend on that of x, then y is said to be a function of two *independent* variables v and x. If we can discover that when x varies y also varies, then y must be a function of x, but this does not hinder that y may not be a function of other quantities as well.

If one quantity can vary while another remains constant, they are *said* to be independent. They are said to be independent, while strictly speaking they are not necessarily so; as, for instance, in the equation

$$\frac{x^2}{a^2} + \frac{y^2}{b^2} + \frac{z^2}{c^2} = 1, \tag{12.8}$$

if we give a definite value (say g) to x, then

$$\frac{y^2}{b^2} + \frac{z^2}{c^2} = 1 - \frac{g^2}{a^2}, \tag{12.9}$$

where we observe that y/b can have any value whatever between two limits:

$$-\sqrt{1 - \frac{g^2}{a^2}} \leq \frac{y}{b} \leq \sqrt{1 - \frac{g^2}{a^2}}.$$

Hence then one quantity may, as regards its general values, be independent of another, while its *limits* are not independent of it.

§3. Again, when a quantity is said to be an independent variable, it is not meant to be implied that it may have any values whatever; for it may happen that the values of a quantity are comprised between certain limits, beyond which it would be absurd to suppose them to extend.

Thus, one sees from Eq. (12.9)

$$\frac{y}{b} = \pm \left(1 - \frac{x^2}{a^2} - \frac{z^2}{c^2}\right)^{1/2};$$
(12.10)

and as the square root of a negative quantity is imaginary,[1] the greatest value of y is $+b$, and the least $-b$. It is capable of any intermediate values whatsoever, but can never pass these values; and, consequently, between these limits we may consider y as an independent variable.

§4. Let u be a function (denoted by f) of an independent variable x; then

$$u = f(x).$$
(12.11)

For x write $x + h$, h being a quantity which is independent of x, but such that $x + h$ is comprised between the limits of the variableness of x; and denote by U the corresponding value of f. We have, therefore,

$$U = f(x + h).$$
(12.12)

We now make the crucial assumption that the right-hand side of Eq. (12.12) can be expanded as a power series in h, so that

$$U = f(x + h)$$
$$= A_0 + A_1 h^a + A_2 h^b + A_3 h^c + \cdots$$
(12.13)

the indices of h being arranged in an ascending order.

In the right-hand side of Eq. (12.13), A_0, A_1, A_2, ...are independent of h, but as the expression is a function of x and h, it is evident that the A_0, A_1, A_2, etc. must be certain functions of x.

§5. The first term A_0 is that part of $f(x + h)$ which does not depend on h for its value, and, consequently, remains the same whatever value be given to h. But if h be made equal to zero, $f(x + h)$ is reduced to $f(x)$ or u, a quantity which is entirely independent of h, and therefore we have always

$$A_0 = u,$$

---

[1] Using the nomenclature of his time, SE calls it *impossible*, not *imaginary*.

$$\therefore U = u + A_1 h^a + A_2 h^b + A_3 h^c + \cdots \qquad (12.14)$$

§6. We observe next that none of the indices of h can be negative. For, if possible, let

$$U = u + A_1 h^a + \ldots \qquad (12.15)$$

Then when $h = 0$, h^{-a} becomes infinite while the other terms of the equation remain finite or vanish, which is impossible.

§7. Nor can any of the indices be fractional. For, if possible, let m/n be a fraction in its lowest terms; and

$$U = u + A_1 h^{-m/n} + A_2 h^b + \cdots \qquad (12.16)$$

Then when $h^{-m/n}$ will have n different values, which we denote by $h_1, h_2, \ldots h_n$,

$$\therefore U_1 = u + A_1 h_1 + A_2 h^b + \cdots ,$$
$$U_2 = u + A_1 h_2 + A_2 h^b + \cdots ,$$
$$\ldots = \ldots\ldots\ldots \qquad (12.17)$$
$$U_n = u + A_1 h_n + A_2 h^b + \cdots , \qquad (12.18)$$

that is, for every value of u, there are n values of U; which is impossible, since the mere writing of $x + h$ for x in any expression cannot possible alter the *number* of forms of x in $f(x)$; consequently, there is only one form of U for each value of u, and, therefore, all the indices of h are whole positive quantities.

§8. Since the quantities u, A_1, A_2, $\ldots$ are independent of h, we may give any values whatever to h without altering them. For h we write $2h$;[2]

$$\therefore f(x + 2h) = u + A_1 2^a h^a + A_2 2^b h^b + \cdots \qquad (12.19)$$

But, if we put $z = x + h$, the left-hand side of Eq. (12.19) equals $f(z + h)$, so that Eq. (12.19) can itself be expressed as

$$f(z + h) = f(z) + \tilde{A}_1 h^a + \tilde{A}_2 h^b + \cdots \qquad (12.20)$$

Since $f(z)$ and the expansion coefficients $\tilde{A}_k$ are functions of $z = x + h$, they can all be expanded as shown below:

---

[2] Very few equations are numbered in the original; Eq. (12.19) is labelled as (A), and Eq. (12.22a) as (B).

$$f(z) = u + A_1h^a + A_2h^b + \ldots$$
$$\tilde{A}_1 = A_1 + B_1h^{a_1} + B_2h^{b_1} + \cdots ,$$
$$\tilde{A}_2 = A_2 + C_1h^{a_2} + C_2h^{b_2} + \cdots ,$$

$$\ldots = \ldots$$

(12.21)

Combining Eqs. (12.21) and (12.20), one gets

$$f(x + 2h) = u + A_1h^a + A_2h^b + \cdots$$
$$+ (A_1 + B_1h^{a_1} + B_2h^{b_1} + \cdots)h^a$$
$$+ (A_2 + C_1h^{a_2} + C_2h^{b_2} + \cdots)h^b + \&c.$$

(12.22a)

$$= u + 2A_1h^a + \text{terms containing } h \text{ to}$$
$$\text{higher powers than } a$$

(12.22b)

Equating the right-hand sides of Eqs. (12.22a) and (12.22b), we get

$$u + A_12^ah^a + A_22^bh^b + \cdots = y + 2A_1h^a + + \cdots$$

(12.23)

Since this equality must always exist, we are led to conclude that

$$A_12^a = 2A_1,$$

(12.24a)

$$\therefore a = 1.$$

(12.24b)

Hence

$$f(x + h) = u + A_1h + A_2h^b + \ldots$$

(12.24c)

§9. The foregoing analysis leads to the following conclusion: In expanding any function of h whatever, the second term will, after having arranged the quantities according to the powers of h, contain h to the first power only; and consequently, in §8, a_1, a_2, ... are each equal to 1. Hence the index of h, which is next greater than a in Eq. (12.22a) is $a + a_1 = 1 + 1 = 2$; but from Eq. (12.19), which is equivalent to Eq. (12.22a), it appears that the index next greater than a is b, which leads us to the conclusion

$$b = 2,$$

(12.25)

which implies, in turn, that $b_1, b_2, \ldots$ are each equal to 2. Again Eq. (12.19) now shows that the index next greater than b is 3, which the equivalent relation Eq. (12.19) shows must be equal to

c; therefore
$$c = 3. \tag{12.26}$$

Similar reasoning shows that the next higher powers of h are 4, 5, ... We are thus led to the expansion

$$f(x+h) = u + A_1h + A_2h^2 + A_3h^3 + \cdots \tag{12.27}$$

§10. The coefficients of the various powers of h, viz. A_1, A_2, A_3 ..., as already observed, are functions of x, and since the whole expression takes its rise from the function derivable from the function $f(x)$, or u; therefore $A_1, A_2, A_3 \ldots$ depend on $f(x)$, and are derivable from it by *particular operations*, independently of h, because their values remain the same whatever be the value of h. The *operation* or *process* of deriving A_1 from f is called *differentiation*, and it (the *process*) will be represented by the symbol D_x; the subscript x, which specifies the *independent variable*, is used for the sake of perspicuity: for as f may happen to be a function of several independent variables $x, y, z, \ldots$ (§2), it would be impossible to ascertain whether x had been changed into $x + h$, or y into $y + h$, or z into $z + h$, unless there were some distinguishing mark for the operation in these distinct cases. By this notation Thus $A_1 = D_x u$.

§11. The nature of the operation denoted by the symbol D_x may be very different for different forms of the function y. It becomes necessary, therefore, to lay down rules for the operation in the several cases that may occur, and this is one object of the science of Differential Calculus; another object is to point out the utility and properties of quantities obtained by differentiation; or, as they are called, *differential coefficients*.

§12. We shall give a few examples of the process of differentiation, that is, of the method of performing the *operation* denoted by D_x.

If $u = f(x)$ and $w = g(x) = Kf(x)$, where K is a quantity which does not depend on x, then

$$g(x+h) = Kf(x+h) = w + KA_1h + KA_2h^2 + \cdots,$$

from which it follows that

$$D_x w = KA_1 = KD_x u. \tag{12.28}$$

Example 1. Suppose $u = f(x)$ to be equal to x^m. Then

$$f(x+h) = (x+h)^m$$
$$= x^m + mx^{m-1}h + \cdots \qquad (12.29)$$

Hence $D_x(x^m) = mx^{m-1}$; that is, the operation of differentiating any power of the independent variable, consists in multiplying by the index, and then subtracting unity from the index.

Example 2. Suppose $u = f(x)$ to e equal to a^x. Then

$$f(x+h) = a^{x+h}$$
$$= a^x \cdot a^h$$
$$= a^x \cdot (1 + \ln a \cdot h + \cdots)$$
$$= a^x + a^x \cdot \ln a \cdot h + \cdots \qquad (12.30)$$
$$\text{Hence } D_x(a^x) = a^x \cdot \ln a \qquad (12.31)$$

Hence $D_x(a^x) = a^x \cdot \ln a$; that is, the operation of differentiating a constant quantity raised to a power whose index is the independent variable, consists in multiplying by the Naperian logarithm of the constant quantity.

§13. *Example 3.* If K be a quantity which is independent of x, then $D_x K = 0$. This may be seen by writing $f(x) = u = Px^0$, and using Eqs. (12.28) and (12.29).

Corollary. Hence $D_x h = 0$. (§4).

§14. To investigate a separate rule of operation for every case that can occur, would evidently be an endless task; and, consequently, methods of reducing the operation in a complex cases to that of a simple one becomes necessary.

Example. To investigate a method of reducing the differentiation of the product of two quantities which are functions of the same independent variable to a more simple case.

Let $u = f(x)$, $v = g(x)$, so that

$$U = f(x+h) = u + D_x u \cdot h + \cdots \qquad (12.32)$$
$$V = g(x+h) = v + D_x v \cdot h + \cdots \qquad (12.33)$$
$$UV = uv + (uD_x v + vD_x u) \cdot h + + \cdots \qquad (12.34)$$

But $UV = f(x+h)g(x+h)$, and is therefore a function of $x+h$, and, consequently, we have

$$D_x(uv) = uD_x v + vD_x u. \qquad (12.35)$$

§15. *Corollary.* If we write w/u instead of v, in the last result, it becomes

$$D_x \left(u \cdot \frac{w}{u} \right) = u D_x \left(\frac{w}{u} \right) + \frac{w}{u} \cdot D_x u, \qquad (12.36)$$

$$\text{or } D_x w = u D_x \left(\frac{w}{u} \right) + \frac{w}{u} \cdot D_x u, \qquad (12.37)$$

$$\therefore D_x \left(\frac{w}{u} \right) = \frac{u D_x w - w D_x u}{u^2} \qquad (12.38)$$

§16. There are many other methods and rules for reducing complex differentiations to simple ones, but as it is not our object to write a treatise on differential calculus, no more examples of that nature need be added here. We shall, therefore, return to the subject as we left it in §10, and since, as was there stated, A_1 is obtained from u by a process called differentiation, and represented by D_x, we shall next inquire by what processes $A_2, A_3, \cdots$ are to be obtained, and how these processes must be represented so as to preserve the uniformity of notation.

§17. We know that

$$f(x+h) = u + D_x u \cdot h + A_2 h^2 + A_3 h^3 + \cdots .$$

If x be changed into $x + k$, k being a quantity independent of x and h, then, as in §8,

$$u \longrightarrow f(x+k) = u + D_x u \cdot k + A_2 k^2 + \cdots \qquad (12.39)$$
$$D_x u \longrightarrow D_x u + D_x(D_x u) \cdot k + B_2 k^2 + \cdots \qquad (12.40)$$
$$A_2 \longrightarrow A_2 + D_x A_2 \cdot k + C_2 k^2 + \cdots \qquad (12.41)$$

and, consequently,

$$
\begin{aligned}
f(x+k+h) = &\; u + D_x u \cdot k + A_2 k^2 + \cdots \\
&+ \{ D_x u + D_x(D_x u) \cdot k + B_2 k^2 + \cdots \} h \\
&\quad + (A_2 + D_x A_2 \cdot k + C_2 k^2 + \cdots) h^2 + \cdots \\
= &\; u + D_x u \cdot h + \cdots + A_2 h^2 + \cdots \\
&+ \{ D_x u + D_x(D_x u) \cdot h + D_x A_2 \cdot h^2 + \cdots \} k + \cdots \qquad (12.42)
\end{aligned}
$$

But by writing $k + h$ for k in the expansion of $f(x+h)$ we have

$$f(x+k+h) = u + D_x u \cdot (h+k) + A_2 (h+k)^2 + A_3 (h+k)^3 + \cdots$$
$$= u + D_x u \cdot (h+k) + A_2 (h^2 + 2hk + \cdots)$$
$$+ A_3 (h^3 + 3h^2 k + \cdots) + \cdots$$
$$= u + D_x u \cdot h + A_2 h^2 + A_3 h^3 + \cdots$$
$$+ (D_x u + 2A_2 \cdot h + 3A_3 \cdot h^2 + \cdots)k + \cdots \qquad (12.43)$$

Since the right-hand sides of Eqs. (12.42) and (12.43) must, like the left-hand sides, be identical, we shall have, by comparing the coefficients of k in the two,

$$D_x u + D_x(D_x u) \cdot h + D_x A_2 \cdot h^2 + D_x A_3 \cdot h^3 + \cdots$$
$$= D_x u + 2A_2 \cdot h + \quad 3A_3 \cdot h^2 + \quad 4A_3 \cdot h^3 + \cdots$$

which equality must exist whatever be the value of h, and therefore, by comparing the coefficients of h, we have

$$2A_2 = D_x(D_x u), \; 3A_3 = D_x(A_2), \; 4A_4 = D_x A_3.$$

Hence $2A_2$ is obtained from $D_x u$, or A_1, by performing the operation of differentiation upon it: $3A_3$ is obtained from A_2; $4A_4$ from A_3, and generally rA_r from A_{r-1} by performing the operation of differentiation in each case.

We *may* conveniently represent $D_x(D_x u)$ by $D_x^2 u$; consequently, $D_x\{D_x(D_x u)\}$, $D_x(D_x^2 u)$ *must* be represented by $D_x^3 u$, and so on. Hence

$$2A_2 = D_x^2 u, \; 3A_3 = D_x A_2 = \tfrac{1}{2} D_x(D_x^2 u) = \tfrac{1}{2} D_x^3 u, \cdots;$$
$$\therefore A_2 = \frac{D_x^2 u}{2!}, \; A_3 = \frac{D_x^3 u}{3!}, \; A_4 = \frac{D_x^4 u}{4!}$$

Wherefore $D_x^r u$ signifies that the process of differentiation is to be performed r times successively, viz.: first on u, then on the quantity resulting; that is, on the differential coefficient of u, then on the result of this operation, and so on, until r differentiations have been performed. Since the result of one differentiation is called the first differential coefficient, $D_x^r u$, which is the result of r differentiations, may be called the r-th differential coefficient.

§18. By substituting for A_2, A_3, etc. their values above determined, we have

$$f(x+h) = u + D_x u \frac{h}{1} + D_x^2 u \frac{h^2}{1 \cdot 2} + D_x^3 u \frac{h^3}{1 \cdot 2 \cdot 3} + \cdots$$

§19. We shall give an example or two, by way of illustrating this formula.

Example 1. To expand $(x+h)^m$.

We know, from principles independent of the binomial theorem, that

$$(x+h)^m = x^m + mx^{m-1} \cdot \frac{h}{1} + \cdots$$

Here $u = x^m$ and $D_x u = mx^{m-1}$; hence the process of differentiation, in this case is the same as that in *Example 1* of §12, and we have

$$D_x u = mx^{m-1}$$
$$D_x^2 u = m(m-1)x^{m-2}$$
$$D_x^3 u = m(m-1)(m-2)x^{m-3}$$
$$\cdots = \cdots$$

Therefore

$$(x+h)^m = x^m + mx^{m-1} \cdot \frac{h}{1} + m(m-1)x^{m-2} \cdot \frac{h}{1 \cdot 2}$$
$$+ m(m-1)(m-2)x^{m-3} \cdot \frac{h}{1 \cdot 2 \cdot 3} + \cdots$$

which is the binomial theorem. We are able, therefore, by merely knowing the process of obtaining the second term from the first, to deduce all the subsequent term from it by the repetition of the same operation.

Example 2. To expand a^h, supposing the second term in the expansion is known to be $\ln a \cdot h$.

$$\text{Here} \qquad a^h = 1 + \ln a \cdot h + \cdots$$
$$\therefore a^{x+h} = a^x \cdot a^h = a^x + a^x \ln a \cdot h + \cdots$$

Hence the process of differentiation is that stated in *Example 2* of §12; and by this process the following quantities are successively obtained from each other:

$$a^x, \ a^x \ln a, \ a^x(\ln a)^2, \ a^x(\ln a)^3, \ \ldots$$

Hence

$$a^{x+h} = a^x + a^x \ln a \cdot \frac{h}{1} + a^x(\ln a)^2 \cdot \frac{h}{2!} + a^x(\ln a)^3 \cdot \frac{h}{3!} + \ldots$$

$$a^h = 1 + \frac{\ln a \cdot h}{1} + \frac{(\ln a)^2 \cdot h^2}{2!} + \frac{(\ln a)^3 \cdot h^3}{3!} + \ldots$$

§20. It appears, therefore, that if we can by any independent means obtain the second term of the expansion of any function of $(x + h)$, we can thence deduce all the subsequent terms of the expansion.

§21. The quantity x, by the variation of which (to $x + h$), u is changed to $f(x + h)$, is called the independent variable (§§2,10): but if x were itself a function of some other quantity t, so as to vary only because t varies, then t, and not x, is said to be the independent variable.

§22. This suggests a very important question; viz., suppose that in any process where *differential coefficients* are employed, we have supposed x to be the independent variable, and then discover that x is itself dependent on another quantity t; what alteration must be made in our result, so as to reduce it to the same form as if we had supposed t to be the independent variable throughout the process? We shall endeavour to solve this question.

§23. On the supposition that x is the independent variable, we have

$$f(x + h) = u + D_x u \cdot \frac{h}{1} + D_x^2 u \cdot \frac{h^2}{1 \cdot 2} + \cdots \tag{12.44}$$

But when t is the independent variable, $x = g(t)$, and suppose that x becomes $x + h$ in consequences of t changing to $t + k$,

$$\therefore x + h = g(t + k) = x + D_t x \cdot \frac{k}{1} + D_t^2 x \cdot \frac{k^2}{1 \cdot 2} + \cdots$$

$$\therefore h = D_t x \cdot k + D_t^2 x \cdot \frac{k^2}{2} + \cdots \tag{12.45}$$

and this value of h, when used in Eq. (12.44), leads, after expanding and arranging the terms in ascending powers of k:

$$f(x+h) = u + D_x u \cdot \left(D_t x \cdot k + D_t^2 x \cdot \frac{k^2}{1 \cdot 2} + \cdots \right) +$$

$$\frac{D_x^2 u}{1 \cdot 2} \left(D_t x \cdot k + D_t^2 x \cdot \frac{k^2}{1 \cdot 2} + \cdots \right)^2 + \cdots$$

$$= u + D_x u \cdot D_t x +$$

$$\left[D_x u \cdot D_t^2 x + (D_t x)^2 \cdot D_x^2 u \right] \frac{k^2}{1 \cdot 2} + \cdots \quad (12.46)$$

But because $x = \phi(t)$, $\therefore u = f(x) = f(g(t)) = \phi(t)$,

$$\therefore f(x+h) = \phi(t+k) = u + D_t u \cdot \frac{k}{1} + D_t^2 u \cdot \frac{k^2}{1 \cdot 2} + \cdots$$

$$\therefore u + D_t u \cdot k + D_t^2 u \cdot \frac{k^2}{1 \cdot 2} + \cdots$$

$$= u + D_x u \cdot D_t x \cdot k + \left[D_x u \cdot D_t^2 x + (D_t x)^2 \cdot D_x^2 u \right] \frac{k^2}{1 \cdot 2}$$

Consequently, by comparing the coefficients of corresponding powers of k, we have

$$D_t u = D_x u \cdot D_t x, \quad \therefore D_x u = \frac{D_t u}{D_t x}$$

$$D_t^2 u = D_x u \cdot D_t^2 x + (D_t x)^2 \cdot D_x^2 u$$

$$= \frac{D_t u}{D_t x} \cdot D_t^2 x + (D_t x)^2 \cdot D_x^2 u$$

$$\therefore D_x^2 u = \frac{D_t x \cdot D_t^2 u - D_t u \cdot D_t^2 x}{(D_t x)^3}$$

$$= \frac{D_t \left(\frac{D_t u}{D_t x} \right)}{D_t x}, \text{ by §15.} \quad (12.47)$$

Similarly, we may find the expressions for $D_x^3 u, D_x^4 u, \ldots$ but they may be deduced more simply from each other thus,

$$\because D_x u - \frac{D_t u}{D_t x}, \quad (12.48)$$

therefore, by writing $D_x u$ for u,

$$\therefore D_x^2 u = \frac{D_t(D_x u)}{D_t x},\qquad (12.49)$$

and, consequently, $D_x^3 u = \dfrac{D_t(D_x^2 u)}{D_t x}$, by writing $D_x u$ for u in Eq. (12.48).

Similarly, $\quad D_x^{m+1} u = \dfrac{D_t(D_x^m u)}{D_t x}. \qquad (12.50)$

§24. Hence the result of any operation performed on the supposition that x is the independent variable, may be changed into an equivalent expression in which t is the independent variable, by substituting instead of $D_x u, D_x^2 u, \dots$ the quantities $\dfrac{D_t u}{D_t x}$, $\dfrac{D_t x \cdot D_t^2 u - D_t u \cdot D_t^2 x}{(D_t x)^3}$, $\dots$respectively. This operation is called "changing the independent variable."

§25. The preceding articles contain the fundamental principles of differential calculus, or, as we will call it in the remainder of the tract, for the sake of distinction, the *calculus of differential coefficients*. We shall now proceed to explain the first principles of what may properly be called the *calculus of differentials*. The two methods have universally been confounded with each other, and have even been called by the name of differential calculus, though their principles, when properly understood, are as perfectly distinct from each other as it is possible for two things to be. Some writers, imagining that the principles of differential calculus were liable to considerable objections, have rejected the science altogether, and adopted the calculus of differential coefficients, but having unfortunately retained the *notation* of the rejected science, which we shall show is peculiar to it, have fallen into great obscurities, and sometimes even into the most palpable contradictions and inaccuracies. Others, again, despising the new science (of differential coefficients), have not even taken the trouble to learn its notation and inquire into its principles, but have condemned it untried as a dangerous novelty. The consequence is, that though nothing is more clear and satisfactory so far as it goes, than the science of the calculus of differentials, when established on proper grounds, and its capabilities defined, yet there is no subject in the whole range of mathematical science, which gives more trouble and is less satisfactory to a learner. The confused and perplexed state of differential calculus may be inferred from the fact, that it would be difficult to find two writers on the subject who agree in the fundamental principles. We state this as an apology for the present production, the purpose of which is to call the serious attention of mathematicians to the subject.

§26. In the method of differentials, the second term of Taylor's expansion of $f(x + h)$ is called the *differential* of the first; the process of obtaining the second term from the first is called *differentiation** and the symbol of this operation is d, consequently the second term is represented by du. Hence the operation denoted by d, when performed on any function of x, produces a quantity which depends on the arbitrary quantity h, (for du consists of a certain function of x multiplied by h); while the operation denoted by D_x, when performed on the same function of x, produces a quantity wholly independent of h; the results of these two operations are therefore perfectly distinct, and the reader must be careful never to confound them.

§27. We have now a new question before us; viz., the second term of the expansion of $f(x + h)$ being derived from the first by an operation called differentiation of which the symbol is d, and being represented by du, by what operation the subsequent terms of that expansion be derived, and how must these operations and terms be represented so as to preserve uniformity of notation.

§28. Since the second term of the expansion of $f(x + h)$ is represented both by du and $D_x u \cdot h$,

$$\therefore du = D_x u \cdot h \tag{12.51}$$

an equality which holds good, whatever be the value of u, providing only it be a function of x. Now, since du depends on $D_x u$, which is a function of x (§10), therefore du is a function of x, and consequently, by writing du for u in the above equation, we obtain

---

*SE's footnote: The same name as used in Art. 10, for a different process; but it is clear a different name ought to be invented for one of them.

184

$$d(du) = D_x(du) \cdot h$$
$$= D_x(D_x u \cdot h) \cdot h \qquad \text{[from Eq. (12.51)]}$$
$$= (D_x^2 u \cdot h + D_x u \cdot h D_x h) h \qquad \text{[by §14]}$$
$$= D_x^2 u \cdot h^2; \qquad \text{[by §13, Cor.]} \quad (12.52)$$

or, since $d(du)$, (which merely signifies that the operation d is to be performed twice) may be more conveniently denoted by d^2u, we may rephrase Eq. (12.52) thus

$$d^2 u = D_x^2 u \cdot h^2. \qquad (12.53)$$

In Eq. (12.53), again, write du for u to get

$$d^2(du) = D_x^2(du) \cdot h^2$$
$$= D_x D_x(du) \cdot h^2$$
$$= D_x(D_x^2 u) \cdot h^3, \text{ as before;}$$

$$\therefore d^3 u = D_x^3 u \cdot h^3 \qquad (12.54)$$

And it is clear that the repetition of the same process would give us

$$d^r u = D_x^r u \cdot h^r \qquad (12.55)$$

Hence the expansion of Taylor, by means of Eqs. (12.51), (12.53), (12.54), ... becomes

$$f(x+h) = u + du + \frac{d^2 u}{1 \cdot 2} + \frac{d^3 u}{1 \cdot 2 \cdot 3} + \cdots \qquad (12.56)$$

The successive terms of Taylor's expansion of $f(x+h)$ are consequently obtained by repeating the process denoted by d, and then dividing by $r!$, where r denotes the number of repetitions.

§29. As in §12, we may investigate rules for finding the differentials of functions of x.

Example 1. Let $u = x^n$,

$$\therefore f(x+h) = (x+h)^n = x^n + nx^{n-1}h + \cdots$$

$$\therefore d(x^n) = x^n + nx^{n-1}h \qquad (12.57)$$

Example 2. Let $u = a^x$,

$$\therefore f(x+h) = a^{x+h} = a^x \ln a \cdot h + \cdots$$
$$\therefore d(a^x) = a^x \ln a \cdot h \qquad (12.58)$$

§30. We may also, as in §14, investigate rules for reducing complex cases to more simple ones.

Example. To find the differential of a product

$$U = f(x+h) = u + du + \cdots$$
$$V = g(x+h) = v + dv + \cdots$$
$$\therefore UV = uv + (udv + vdu) + \cdots$$
$$\therefore d(uv) = udv + vdu \qquad (12.59)$$

§31. *Corollary.* For v write $\dfrac{w}{u}$ in the last equation (w being a function of x),

$$\therefore d\left(u \cdot \frac{w}{u}\right) = ud\left(\frac{w}{u}\right) + \frac{w}{u} \cdot du,$$
$$\text{or } dw = ud\left(\frac{w}{u}\right) + \frac{wdu}{u};$$
$$\therefore d\left(\frac{w}{u}\right) = \frac{udw - wdu}{u^2}. \qquad (12.60)$$

§32. We shall now consider a similar question to that of §22. But first, we must observe that du is obtained from u by performing upon it some operations with regard to x, of which it is a function, it is necessary when u is a function of several independent variables $x, y, z \ldots$ to know, with regard to which of them the operation d is to be performed; for the results may be very different, according as d is performed with regard to the one or the other. And herein the notation of differentials is defective, for when we meet with the expression du, it is impossible to know what it represents. We know, indeed, that an operation of a certain kind is to be performed on u in respect of some quantity of which u is a function, but which is that quantity, it is impossible to tell. Hence arise all that confusion and obscurity from which

186

very few, if any treatises on differentials are free; and, in this respect also, it is very much inferior to the calculus of differential
coefficients, which is remarkable for its perspicuity. Whenever
therefore we meet with the expression du, since there is no more
reason why the operation d is to be performed with respect to one
variable than another, we must suppose it to be performed with
regard to all the variables of which u is a function, these variables
being determined as in §1.

§33. To proceed, however, with the question under consideration, viz.: suppose that in any process where differentials are
employed, x has been taken for the independent variable, and
it is discovered that x is itself a function of another quantity t,
which is consequently the independent variable; what alteration
must be made in our result, so as to reduce it to the same form as
if t had been supposed the independent variable throughout the
process?

In this article u is supposed a function of *one* independent
variable only; and the operation of finding the differential will
be denoted by d when x is considered the independent variable;
and by δ when t is considered the independent variable. Suppose
that it is in consequence of t becoming $t + k$ that x becomes $x + h$.

By §28 we have

$$du = D_x u \cdot h, \quad d^2 u = D_x^2 u \cdot h^2, \quad d^3 u = D_x^3 u \cdot h^3, \cdots \qquad (12.61)$$

Similarly

$$\delta u = D_t u \cdot k, \quad \delta^2 u = D_t^2 u \cdot k^2, \quad \delta^3 u = D_t^3 u \cdot k^3, \cdots \qquad (12.62)$$

and

$$\delta x = D_t x \cdot k, \quad \delta^2 x = D_t^2 x \cdot k^2, \quad \delta^3 x = D_t^3 x \cdot k^3, \cdots \qquad (12.63)$$

Now, by §23,

$$h = D_t x \cdot k + D_t^2 x \cdot \frac{k^2}{1 \cdot 2} + D_t^3 x \cdot \frac{k^3}{1 \cdot 2 \cdot 3} + \cdots$$
$$= \delta x + \frac{\delta^2 x}{1 \cdot 2} + \frac{\delta^3 x}{1 \cdot 2 \cdot 3} + \cdots ; \qquad (12.64)$$

from which equation h is known.

And from the same article

$$D_x u = \frac{D_t u}{D_t x},$$

$$\therefore D_x u \cdot h = \frac{D_t u \cdot k}{D_t x \cdot k} \cdot h,$$

$$\therefore du = \frac{\delta u}{\delta x} \cdot h \tag{12.65}$$

Also, because $D_x^2 u = \dfrac{D_t x D_t^2 u - D_t u D_t^2 x}{(D_t x)^3}$

$$\therefore D_x^2 u \cdot h^2 = \frac{D_t x \cdot k \cdot D_t^2 u \cdot k^2 - D_t u \cdot k \cdot D_t^2 x \cdot k^2}{(D_t x \cdot k)^3} \cdot h^2, \tag{12.66}$$

$$\therefore d^2 u = \frac{\delta x \delta^2 u - \delta u \delta^2 x}{(D_t x)^3} \tag{12.67}$$

or, as in §13, $\qquad d^2 u = \dfrac{\delta(D_x u)}{\delta x} \cdot h^2, \tag{12.68}$

and, similarly, $\quad d^{m+1} u = \dfrac{\delta(D_x^m u)}{\delta x} \cdot h^{m+1};$

$$\tag{12.69}$$

h is to be eliminated from each of these expressions, if necessary, by means of Eq. (12.64). Any result containing differentials with regard to x, may be reduced to an equivalent result containing only differentials with regard to t, by means of Eqs. (12.64), (12.66), (12.67) ....

§34. In the preceding articles, beginning with §25, the principles of the calculus of differentials have been considered, and to us appear perfectly satisfactory, though the notation, when functions of several independent variables are concerned, is somewhat defective in neatness, and certainly very much inferior to that of calculus of differential coefficients. We shall now proceed to consider some cases, where the notations of both may be advantageously employed together.

§35. The quantity u being a function of two independent variables x and y, it is required to find its differential.

Here $u = f(x,y)$, and by the theorem of §28, using d_x as the symbol of differentiation with regard to x (see §32), we have

$$f(x+h,y) = u + d_x u + \frac{d_x^2 u}{1 \cdot 2} + \cdots \tag{12.70}$$

and denoting $f(x+h,y)$ by u_1, and using d_y as the symbol of differentiation with respect to y, we have, by the same theorem,

$$f(x+h,y+k) = u_1 + \frac{d_y u_1}{1} + \frac{d_y^2 u}{1 \cdot 2} + \cdots$$

But because $u_1 = u + \dfrac{d_x u_1}{1} + \dfrac{d_x^2 u}{1 \cdot 2} + \cdots$

$$\therefore \quad d_y u_1 = d_y u + \frac{d_y d_x u}{1} + \cdots$$

and $\quad d_y^2 u_1 = d_y^2 u + \cdots$

$$\therefore f(x+h,y+k) = u + d_x u + \frac{d_x^2 u}{1 \cdot 2} + \cdots$$
$$+ d_y u + d_y d_x u + \cdots$$
$$+ \frac{d_y^2 u}{1 \cdot 2} + \cdots$$
$$= u + (d_x u + d_y u) + \frac{1}{1 \cdot 2}(d_x^2 u + 2 d_y d_x u + d_y^2 u) + \cdots$$

Hence
$$du = d_x u + d_y u, \tag{12.71}$$
$$d^2 u = d_x^2 u + 2 d_y d_x u + d_y^2 u, \tag{12.72}$$
$$\&c. = \&c.$$

§36. *Corollary 1.* If u were a function of any number of independent variables $x, y, z \ldots$ we should, in like manner, find

$$du = d_x u + d_y u + d_z u + \cdots \tag{12.73}$$

§37. *Corollary 2.* When a quantity, which is a function of several independent variables, is differentiated with regard to one only, the result is called a *partial differential*.

It is evident from Eq. (12.73) that the differential of any quantity is equal to the sum of all its partial differentials.

§38. *Corollary 3.* It is quite evident that besides Eq. (12.72), we must have

$$d^2u = d_x^2u + 2d_xd_yu + d_y^2u, \tag{12.74}$$

and, therefore,

$$d_x^2u + 2d_yud_xu + d_y^2u = d_x^2u + 2d_xd_yu + d_y^2u,$$
$$\therefore d_yd_xu = d_xd_yu. \tag{12.75}$$

It follows from the above that when we have to differentiate successively with regard to several independent variables, it is immaterial in what order (as regards the variables) the differentiations be performed.

§39. From the above specimen, where only two independent variables are concerned, it is manifest how clumsy and inconvenient the notation of differentials becomes when several variables are concerned. It becomes necessary, therefore, to invent some more convenient method.

There is no theorem corresponding to that of §37 in the calculus of differential coefficients; but

since $d_xu = \mathscr{D}_xu \cdot h,$ and $d_yu = \mathscr{D}_yu \cdot k$ [§28] (12.76)

$\therefore d_y(d_xu) = \mathscr{D}_y(\mathscr{D}_xu \cdot h) \cdot k$ and $d_x(d_yu) = \mathscr{D}_x(\mathscr{D}_yu \cdot k) \cdot h,$

or $d_yd_xu = \mathscr{D}_y\mathscr{D}_xu \cdot hk,$ and $d_xd_yu = \mathscr{D}_x\mathscr{D}_yu \cdot kh;$

$\therefore$ $\mathscr{D}_y\mathscr{D}_xu \cdot hk = \mathscr{D}_x\mathscr{D}_yu \cdot kh,$

$\therefore$ $\mathscr{D}_y\mathscr{D}_xu = \mathscr{D}_x\mathscr{D}_yu, \tag{12.77}$

which exactly corresponds to Eq. (12.75).

§40. Again, substituting in Eq. (12.73), we have, by means of §28,

$$du = \mathscr{D}_xu \cdot h + \mathscr{D}_yu \cdot k + \mathscr{D}_zu \cdot l + \cdots \tag{12.78}$$

a theorem, which is exactly equivalent to that of §36, and in which there is no ambiguity or clumsiness.

By substituting in Eq. (12.72), we have, by means of §28,

$$d^2u = \mathscr{D}_x^2u \cdot h^2 + 2\mathscr{D}_x\mathscr{D}_yu \cdot hk + \mathscr{D}_y^2u \cdot k^2, \tag{12.79}$$

which is also remarkable for its perspicuity and neatness.

§41. Before leaving the subject we may just observe, that in order to preserve uniformity of notation, the quantities $h, k, l, \ldots$ must be respectively represented by $dx, dy, dz \ldots$ but the reader must keep in mind that though x appears in the expression dx, yet dx is a quantity entirely independent of x, similarly dy and dz are independent of y and z.

To prove that $h = dx$, let $f(x) = x = x^1$,

$$f(x + h) = (x + h)^1 = x + h + 0 \cdot h^2 + \cdots$$
$$\therefore dx = h.$$

Hence the theorems before mentioned become still more simple and elegant, for by substituting for h, k, &c.

$$du = \mathscr{D}_x u \cdot dx + \mathscr{D}_y u \cdot dy$$
$$d^2 u = \mathscr{D}_x^2 u \cdot dx\, dx + 2\mathscr{D}_x \mathscr{D}_y u \cdot dx\, dy + \mathscr{D}_y^2 u \cdot dy\, dy,$$
$$\&c. = \&c.$$

and whatever be the number of variables,

$$du = \mathscr{D}_x u \cdot dx + \mathscr{D}_y u \cdot dy + \mathscr{D}_z u \cdot dz + \cdots \tag{12.80}$$

It is impossible to conceive a more *perfect* notation than this.

§42. By §28, where u is a function of *one* independent variable,

$$du = D_x u \cdot h, \quad d^2 u = D_x^2 u \cdot h^2, \quad d^3 u = D_x^3 u \cdot h^3 \tag{12.81}$$
$$= D_x u \cdot dx, \quad = D_x^2 u \cdot dx^2, \quad = D_x^3 u \cdot dx^3 \tag{12.82}$$
$$\therefore D_x u = \frac{du}{dx}, \quad D_x^2 u = \frac{d^2 u}{dx^2}, \quad D_x^3 u = \frac{d^2 u}{dx^3}, \tag{12.83}$$

Whenever, therefore, u is a function of one independent variable, the differential coefficients *may* be represented by the fractions

$$\frac{du}{dx}, \quad \frac{d^2 u}{dx^2}, \quad \frac{d^3 u}{dx^3}, \ldots \tag{12.84}$$

But if u be a function of two independent variables x and y,

then, because $\qquad\qquad du = \mathscr{D}_x u \cdot dx + \mathscr{D}_y u \cdot dy, \qquad (12.85)$

we have $\qquad\qquad \dfrac{du}{dt} \equiv D_t u = \mathscr{D}_x u \dfrac{dx}{dt} + \mathscr{D}_y u \cdot \dfrac{dy}{dt}, \qquad (12.86)$

$$= \mathcal{D}_x u D_t x + \mathcal{D}_y u \cdot D_t y \qquad (12.87)$$

$$\text{and} \therefore \text{with } t = x, \quad D_x u = \mathcal{D}_x u + \mathcal{D}_y u \cdot \frac{dy}{dx}, \qquad (12.88)$$

an important result to which we shall return, but we will first say a few words about the rest of Earnshaw's pamphlet.

Earnshaw's error

The equation that corresponds to Eq. (12.85) appears as $du = d_x u \cdot dx + d_y u \cdot dy$ in Earnshaw's pamphlet; as explained earlier, he used the same symbol (a subscripted d) for representing both partial and total derivatives with respect to the subscript, thereby vitiating the rest of his argument. I quote (verbatim, apart from replacing d_x by D_x), the counterpart of Eq. (12.85) and the next equation, along with a short excerpt of the following text:

"then, because $\quad du = D_x u \cdot dx + D_y u \cdot dy \qquad (12.89)$

we have $\quad D_x u = \dfrac{du}{dx} - D_y u \cdot \dfrac{dy}{dx}, \qquad (12.90)$

from which it appears that the differential coefficient $D_x u$ can no longer be represented by the fraction $\dfrac{du}{dx}$. Hence the only ..."

This is where p. 24 of the pamphlet ends; the contents of the last two pages need not be reproduced here because most of it will be found in two critical reviews of the pamphlet, one by De Morgan (pp. 192–204), the other (pp. 209–223) by an author whose name was given as "J. J.".

Notation for partial differentiation

It will be instructive to see how Hardy states the contents of Eq. (12.85). He considers a function $f(x,y)$, with $y = \psi(x)$ and writes [59, p. 178]:

We then obtain

$$D_x f\{x, \psi(x)\} = D_x f(x,y) + D_y f(x,y)\, \psi'(x), \qquad \text{(GHH-1)}$$

where y is to be replaced by $\psi(x)$ after differentiation.

It was this case which led to the introduction of the notation $\partial f/\partial x, \partial f/\partial y$. For it would seem natural to use the notation df/dx for either of the functions $D_x f\{x, \psi(x)\}$ and $D_x f(x, y)$ in one of which y is put equal to $\psi(x)$ before and in the other after differentiation. Suppose for example that $y = 1 - x$ and $f(x, y) = x + y$. Then $D_x f(x, 1 - x) = D_x 1 = 0$, but $D_x f(x, y) = 1$.

The distinction between the two functions is adequately shown by denoting the first by df/dx and the second by $\partial f/\partial x$, in which case the theorem takes the form

$$\frac{df}{dx} = \frac{\partial f}{\partial x} + \frac{\partial f}{\partial y}\frac{dy}{dx};$$ (Hardy-2)

though this notation is also open to objection, in that it is a little misleading to denote the functions $f\{x, \psi(x)\}$ and $f(x, y)$ whose forms as functions of x are quite different from one another, by the same letter f in df/dx and $\partial f/\partial y$.

We see again that it is not absolutely necessary to use different symbols (D and $\mathscr{D}$) for distinguishing partial from ordinary derivatives. Hardy decides to use only a subscripted D, but then he has to state the arguments of the differentiated function to enable the reader to be able to infer that that $D_x f\{x, \psi(x)\}$ is an ordinary derivative, whereas $D_x f(x, y)$ and $D_y f(x, y)$ are partial derivatives. Similarly, we can dispense with ∂ in Eq. (Hardy-2) and express in the following form

$$\frac{df}{dx} = \left(\frac{df}{dx}\right)_y + \left(\frac{df}{dy}\right)_x \frac{dy}{dx},$$ (12.91)

but it must be admitted that no particular advantage results from using this form.

Let us return now to Earnshaw's tract and read an anonymous review [162] whose author is now known to have been Augustus De Morgan.

De Morgan's critique of Earnshaw's tract

So long as the notation which forms at the subject of this article remained confined to a few elementary works, there was little need for any one who valued the high and increasing scientific

character of the University, to do more than regret the unnecessary trouble which would be imposed upon those learners, who
should, after being trained in the new system, attempt to read the
works of Lagrange, Laplace, or Poisson. But the occurrence of the
new language in the public examination papers for the present
year, as given in the Cambridge Calendar, has clothed the innovation with another character; and, though we confess we have
not much fear, has led us to think there may possibly be danger of
the harmony being interrupted which at present exists between
the English and foreign scientific language. We cannot abstain
from offering some observations upon the inconvenience which
would result from such a change; but, to avoid any particular remarks, we shall not mention even the names of the authors to
whom we refer. They are very worthy sons of their *Alma mater*,
evidently masters of their subject, but led aside on this particular
point, as appears to us, not by any false mathematical views, but
by a wrong estimate of the balance of convenience and inconvenience.

The simultaneous invention of the theory of fluxions by
Newton, and of the differential calculus by Leibnitz, was rich in
useful consequences, by the various lights in which it caused every question to be viewed. But the difference of notation was
an evil which, to this country, more than counterbalanced all
the advantages. The whole continent adopted the symbols of
Leibnitz; the English retained those of Newton, and gradually
lost their mathematical character. The reason is obvious enough:
our neighbours, with a more general and powerful notation,
could easily translate all that was done in England; while we,
on the contrary, could not, without great difficulty, make the language of fluxions tell us all that was discovered abroad. They
had also the advantage of numbers and international communication; we could hardly read their writings, and could not, or
at least did not, introduce their new and powerful methods of
investigation. And to increase the difficulty, any attempt at innovation was considered as a sin against the memory of Newton.

The University of Cambridge first broke through the mist
which hid the whole continent from our view. In 1803, Mr. Woodhouse published his *Principles of Analytical Calculation*, in which
the notation of Leibnitz was explained and dwelt upon. The im-

pulse was thus given, though not with very great force. In 1813, it appeared from the *Memoirs of the Analytical Society*, a body of juniors, among whom were Messrs. Herschel, Peacock and Babbage, that the change had several zealous advocates. In 1816, these gentlemen published a translation of Lacroix's Differential Calculus, with a volume of examples, and in 1817, the second named introduced the *Differential Calculus* formally into the public examinations. Since that time the new system must be considered as established.

That the differential notation possessed many advantages over the fluxional was soon almost universally admitted. But this, in our opinion, would alone have been but a poor defence of the change. Had the question been upon a matter of reasoning, we need hardly say, that the united opinion of European mathematicians should not have influenced an examiner, further than to induce him to look very narrowly, and be more than commonly sure of his ground, for his own sake, before he ventured to differ from so large a majority of thinking men. But on a question of language, that of the majority is in most cases the best, because it is that of the majority. It is essential that the scientific communication of all countries should be as open as possible; and we feel very sure, that had the notation of fluxions prevailed over the continent at the period of which we have been speaking, no one of those who were instrumental in promoting the Cambridge reformation would have called it by that name, or put his hand to the work.

And it must be observed, that the change which has produced such beneficial consequences appears, at the first glance, a very trifling one. We write the notation of fluxions and differentials in a few cases:

fluxions	$\dfrac{\dot{y}}{\dot{x}}$	$\dfrac{\ddot{y}}{\dot{x}^2}$	$\dfrac{\ddot{y}}{\dot{x}\,\dot{z}}$	$\dfrac{\dddot{y}}{\dot{x}^2\,\dot{z}}$
differentials	$\dfrac{dy}{dx}$	$\dfrac{d^2y}{dx^2}$	$\dfrac{d^2y}{dx\,dz}$	$\dfrac{d^3y}{dx^2\,dz}$

It should seem then that the two notations may be reduced to perfect coincidence, by so slight a convention as agreeing that a dot above a letter shall mean that *d* is to be read before it; which does not apparently require more effort than to recollect, that in old printing of Latin works, a circumflex above a vowel means that m is to be read after it. But this little difference widened the straits of Dover many thousand times, so far as the mathematical sciences were concerned; which is a very strong argument against any variation, however small, from universal practice. At the same time it serves to show, that those who are determined to innovate, may as well at once carry their system to the very furthest point their fancy or judgment will let them go.

Suppose that the inhabitants of continental Europe had, by some circumstance or other, come to speak a common tongue—that the English, sensible of the facilities which their accession to the same would procure for themselves, had, with great pains, educated an entire generation in the new language, so that the old had been entirely thrown aside: what should we say to an attempt to invent and introduce a third, which had never, till then, existed in any country, simply because the framers thought their own invention more expressive or more beautiful? We should laugh, and say they never could succeed; but unfortunately, the case before us, to which the preceding bears great analogy, is one in which they *may* succeed, if not checked in time. And unless it is to be granted that Cambridge can preserve its scientific reputation and utility by itself, without any communication with foreign, or even with the rest of British science, it behoves those whose influence may be of use, to express their opinion on the subject. We have elsewhere (vol. iii. p. 276) advocated that part of the University system, which consists in non-interference with college tuition, and we should deprecate any public attempt, on the part of the whole corporation, either to aid or obstruct anything of the kind. Discussion and example must be the influences employed, and we hope that there are many in the University, whose authority will have weight, and who will do one of two things: either openly declare themselves in favour of the new system, to the end that men of science throughout Europe may know by how great an authority the change is recommended, and may be induced to consider the propriety of imitation; or at once to oppose a measure which cannot be of little consequence either way, but must be either very good or very bad. Neutrality on their parts will lead to an inference—either that they are indifferent to the University system, which is not true—or that they despise the attempt and those who make it, which ought not to be true, for the change certainly comes upon us recommended by men of talent, and busily employed in writing elementary works for the use of the students in the University.

The new notation is one to which no objection could have been advanced a hundred years ago, though we cannot but declare that we think we should, even upon an unprejudiced examination, have preferred that of Leibnitz. It will be best understood by making a comparison of the two.

$$
\begin{array}{c c c c c}
\text{Leibnitz} & \dfrac{dy}{dx} & \dfrac{dy}{dz} & \dfrac{d^2y}{dx^2} & \dfrac{d^2y}{dx\,dz} \\[2ex]
\text{New Cambridge System} & d_x y & d_z y & d_x^2 y & d_x\,d_z y
\end{array}
$$

We now come to the arguments adduced in favour of the change, which we take from a work entitled *On the Notation of the Differential Calculus*: Deighton, Cambridge, 1832. On looking through the tract, we see with some satisfaction, that its intelligent author does not seem to feel quite sure the University will see exactly where the beauties of his system lie, unless he points out where and what they are to admire. If there be anything that carries conviction in one moment to the mind, it is a real improvement of mathematical notation. The successive authors who brought a^6 into its present form through the following stages; *a cubo-quadratum aaaaaa*; a^6; never needed even a note of exclamation to call the attention of their readers to the convenience of each step. But in the tract be fore us, we are shown where we are to find 'no ambiguity or clumsiness,'—where there is 'perspicuity and neatness,'—where the latter becomes 'still more simple and elegant,'—where 'it is impossible to conceive a more *perfect* notation.' And as to the system of Leibnitz we are told, for fear we should not find out—where it is 'inconsistent with itself,'—where 'ridiculous subterfuges' have been used—and where it is 'a matter of wonder that that notation has not been long ago banished.' All this we see with satisfaction, for we always suspect the weak point where we see most parade of attack and defence. We shall now, without a word of either admiration or censure, proceed to state why we think the new system not preferable to that of Leibnitz on any point, and inferior to it in several. Firstly, the weak points of the notation of Leibnitz seem to us to be faithfully preserved in the new system. Let us take an example from our tract:—

'z being a function of x and y, two independent quantities, the following equation is said to express the connexion between $\dfrac{dz}{dx}$ and $\dfrac{dz}{dy}$

$$\frac{dz}{dx} = \frac{dz}{dx} + \frac{dz}{dy} \cdot \frac{dy}{dx} \qquad (1)$$

from which we should naturally conclude that

$$0 = \frac{dz}{dy} \cdot \frac{dy}{dx}$$

Now we ask what connexion does this establish between $\dfrac{dz}{dx}$ and $\dfrac{dz}{dy}$? —Certainly none. In order however to explain how equation (1) does represent such a connexion, we are told that "$\dfrac{dz}{dy}$ on the *left* hand side of the equation does not mean the same thing as $\dfrac{dz}{dy}$ on the right hand side," an explanation not likely to be very satisfactory to a learner.'

This instance is taken from a work 'the title of which,' says the author, 'it is not necessary to mention.' Here we differ: we think it was *very* necessary to mention the author of the work, that the opponents of the new system might know whether he was one whom they would accept as a fair stater of their case. If any author of reputation has made such an explanation, which may be the case, surely the author of our tract is aware that ninety-nine out of a hundred would not follow him. Has the author of the tract never seen

$$\frac{d.z}{dx} \qquad \frac{d(z)}{dx} \qquad \frac{d}{dx}z \qquad \text{or} \qquad \frac{1}{dx}dz$$

purposely employed to distinguish the $\dfrac{dz}{dx}$ on the first side from that on the second? Let him consult the Cambridge translation of Lacroix, page 158, where he will see $\dfrac{du}{dx} + \dfrac{du}{dy}\dfrac{dy}{dx}$ represented, not by $\dfrac{du}{dx}$, but by $\dfrac{d(u)}{dx}$.

But this is not our strongest objection to the paragraph cited. We see in it that the author has not chosen to bring the new system into juxtaposition even with his own unfair specimen of the old. On looking through the tract we cannot find a single reservation which should hinder us from supposing that his own method of writing the equation (1) would be

$$d_x z = d_x z + d_y z \, d_x y$$

which is liable to his own objection. If he prefer

$$d_x(z) = d_x z + d_y z \, d_x y$$

he does no more than, as he ought to have stated, is *almost* (but for his own quotation, as far as we know, *quite*) universally done in the old system.

Secondly, our author remarks :—

'At page 175 of the Cambridge translation of *Lacroix*'s Differential and Integral Calculus, we have these two equations,

$$\frac{du}{dx} + \frac{du}{dz}\cdot\frac{dz}{dx} = 0 \quad \text{and} \quad \frac{du}{dy} + \frac{du}{dz}\cdot\frac{dz}{dy} = 0$$

on which the author remarks, "the dz of the first equation must not be *confounded* with that of the second." Now we ask what is there to distinguish dz in the one, from dz in the other?—Nothing. In fact, this remark alone ought to have been sufficient to demonstrate the necessity of an improvement in the notation. A little below, in the same page, we find the two following *explanatory* equations,

$$dz = \frac{dz}{dx}\,dx \quad \text{and} \quad \frac{dz}{dy}\,dy$$

which we hold to be utterly *unintelligible*, though they are given by way of *explaining* the mystery of their predecessors above.'

On this, we first beg to assure the author, that we do not mean to charge him with intentional unfairness. The more than common evidence of carelessness in his quotation saves him from any such imputation. He is labouring under a prepossession which prevents him from reading correctly anything written in the notation of Leibnitz: in proof of which, we give the whole passage from the Cambridge translation, recommending him strongly to be very careful in future how he quotes.

'The dz of the first equation must not be confounded with that of the second; for they are both only partial differentials, as has been remarked in No. 120: the complete differential, which is the sum of the terms of the first order, is

$$dz = \frac{dz}{dx}\,dx + \frac{dz}{dy}\,dy = p\,dx + q\,dy$$

p representing $\frac{dz}{dx}$ and q representing $\frac{dz}{dy}$.

When we have simply $dz = p\,dx$, dz is the differential of the ordinate of the section parallel to the plane of x and z ; and similarly $dz = q\,dy$ is that of the ordinate of the section parallel to the plane of y and z.'

Where are the *explanatory* equations? We cannot find them in any part of the chapter from which he quotes. The author should, in fairness, have carried his first quotation as far as 'No. 120,' in the preceding: for there, and there the student is told to refer, would be found,

'In general, when a function of several variables is concerned, it should be remembered, that in $\frac{dz}{dx}$, dz is the partial differential of z relative to x, whilst in $\frac{dz}{dy}$, dz is the partial differential relative to y.'

We are not inclined to admit, with the author, that there is *nothing* to distinguish dz in the first equation from dz in the second, when it is most specifically laid down that the first is never to appear except over dx, or the second except over dy. A lame distinction he may think it, but surely it *is* a distinction. Would he say that there was nothing to distinguish two perfectly similar horses, when it was always agreed and kept to, that one should wear a white saddle, and the other a black one? The saddle might certainly be put on the wrong horse, or it might be that some spectators would be willing to swear black was white, but what argument can be drawn from that against $\frac{dz}{dx}$ and $\frac{dz}{dy}$ which does not more strongly apply to $d_x z$ and $d_y z$, on account of the diminished size of the sole distinction, the x and y?

The only argument against the usual form is, that from $\frac{dz}{dx} = p$ it is not allowable to deduce $dz = pdx$ because then dz, being reft of that distinction which explains what dz it is, may be mistaken for another dz. The answer to this is, that for that very reason $dz = pdx$ never *is* deduced, or at least, if, on some rare occasion, it should be necessary, either special attention is paid not to make any mistake, or, which is better, the equation is written $\frac{dz}{dx} = p \cdot dx$; so that the algebraical multiplication of $\frac{dz}{dx}$ and dx is denoted, but not reduced to its simplest form. Far from thinking this a 'ridiculous subterfuge,' our opinion is, that it is one of the most happy refinements which have attended the progress of mathematical language: and we judge it, not by its appearance, but by the actual consequences which have followed from its use, and the great practical facilities which we have found it afford. At the same time $d_x z$ is in no ways inferior, so far as we have yet considered the subject: and this brings us from the negation of inferiority in the old system to those points on which we assert its superiority.

The system of Leibnitz is superior to the new one in legibility. It will be admitted that x in $\frac{dz}{dx}$ is much more easily read than in $d_x z$; and this becomes of consequence, when it is considered that the x is the sole distinction in both cases between one dz and another. But this, though of considerable importance, does not require dwelling upon.

The new system commits breaches of analogy which are not to be found in the old one. We quote an instance from our tract.
$$du = d_x u \cdot dx + d_y u \cdot dy + d_z u \cdot dz + \cdots$$
'It is impossible to conceive a more perfect notation than this.' Perfectly possible, we reply, and for the following reasons. In mathematical language, it is desirable that things which are very

unlike in character should have symbols which are also very un-like in form. Now, there is great likeness between du and $d_x u$; a small omission of the pen, or too quick a glance of the eye, would confound them. But du is an increment which may be made as small as we please, and of which it is repeatedly necessary to suppose that it should diminish without limit. But $d_x u$ is, in the theory of limits, not a diminishing variable, but the ratio of two diminishing variables, which ratio cannot be made to diminish without limit, except in certain cases: in the theory of expansions it is the algebraical co-efficient of a development. We must then put this strongly, that considering the very great difference be-tween the meanings of du and $d_x u$, there is not sufficient differ-ence between the symbols.

Again, in every other case of indices, whether written above or below the letter, it so happens that a and a [a_1?] are so con-nected, that a is a_1 with the figure omitted. Thus when $n = 1$ $a^n = a^1$ which is generally written a; and in the series of coeffi-cients a_1 a_2 a_3 &c., the first is frequently written a, though very often the breach of analogy a, a_1, a_2 &c., is committed. But even in the latter case a here stands for a_0, and if a_n be a function of n, is what a_n becomes when $n = 0$. Let us now apply the same ex-tension to $d_x u$: in this case, du ought to stand for the differential co-efficient of u with respect to *one*; an absurdity, since one is not a variable. We must, therefore, protest against $d_x u$, as a breach of analogy, which may hereafter stop the progress of discovery. Many of our readers must know how much the perfect symmetry of algebraical language has hitherto aided investigation.

If the predominant idea in the mind of the mathematician, when he sees the letter d, is not to be—that the letter which fol-lows it has received an increment, which increment is either in-finitely small, in the incorrect language of one system, or allowed to be diminished to any extent, in that of another—if such is not to be his notion, he must not read the mathematical works of Euler, D'Alembert, Clairaut, Legendre, Lagrange, Laplace, Pois-son, Cauchy, &c., &c., or those of our most celebrated living En-glish authors. For he cannot read these without retaining an idea which runs through them all. In truth, we believe that, in the closet, the mathematician, in nine cases out of ten, speaks to him-self in the language of infinitely small quantities. This theory has the advantage of a peculiar simplicity, from and after the first step, which is erroneous. But the error is one rather of language than of principle, in the opinion of many: and it is certain that it can always be corrected at any part of the process, requiring only circumlocution to make the very same step with the very same symbols, in a manner perfectly logical. He who uses infinitely small quantities says that x can be taken so small that $a + x$ shall be $= a$; which is certainly not true. In the theory of limits, it is said that x can be taken so small that $a + x$ shall be, as nearly as we please, $= a$. The first says that the ratio of $a + x$ to a can be absolutely 1; the second, that the limiting ratio of $a + x$ to a, or the ratio which that of $a + x$ to a continually approaches, but never reaches, so long as x has any value, is 1.

When two systems prevail, which are so like each other, that the second may be, and often is, considered as a derivation from the first, made by dropping a cumbrous circumlocution, which is necessary in logic, but not in practice, when once well under-stood, it is most desirable that the notations of the two systems should be the same. This is the case in the common notation. The only distinction is, that whereas an equation of the form $dy = pdx$ can be written in the second, and is not strictly true, but can be brought as near the truth as we please by diminishing dx, this equation is writtten in the first $\dfrac{dy}{dx} = p$ and is strictly true, for this reason, that it is agreed the first side shall not represent the ratio of the increment of y to that of x but the *limit* of that ratio.

But if it is still considered necessary to have some *elementary symbol* for the differential coefficient, to be used when more convenient than that just described, which would have its ad-vantages in certain cases, why take the letter d, which has ac-quired throughout Europe a permanence of meaning unattained as yet by any mathematical symbol whatsoever? Surely the let-ter D might serve the purpose as well, and would be as easy to write as d. The old notation might thus be entirely preserved, and $dy = D_{x}y \cdot dx$ would be immediately understood by the partisans prefers of every notation. And there would be the additional advantage D_x to d_x of writing the independent variable inside the D, and in time, we doubt not, of printing it in the same way. This would render any breach of analogy impossible, since the position is perfectly new.

But without being as positive as the author of our tract, we will not say it is impossible to conceive a more *perfect* notation than that of Leibnitz, but only that we do not believe any one which combines so many advantages has yet been brought forward: and that, even could a system be invented which should unite all suffrages, it would be unwise to conclude that we should gain by adopting it, in the face of the consideration, that nearly all the works which have been written in the century and a half which has been most productive in mathematical discovery, would be almost unreadable by the next generation.

Most, all we believe, of the Cambridge works which advocates the *entire* new system have emanated from St. John's College; but the first step was made in Trinity College. Most unfortunately, a very eminent ornament of the University was incautious enough to tell the world by his example, that he considered the objections of an individual to a universally received notation were sufficient reason for making use of one of his own, without reflecting upon the inconveniences which would result, if every one were to adopt the same course. His change was the writing $\int_x y$ for $\int y \, dx$, of which the subject of this article is only an extension. But had he remembered that another might object to the differential notation, a third to some other part, and so on, we think he would hardly have made himself in any degree an advocate for individual license in matters of notation, except upon points which have not been settled by custom. A pretty system of mathematics we should have, if every man arranged all the symbols according to his own judgment of what would have been best, had he been to begin *de novo*; and a comfortable prospect for the beginner!

But if the stream cannot be stemmed, we mean, for our own parts, to sail even faster than the current. We dream already of teaching—that the places of the root and exponent ought to be reversed,—that + does better to denote subtraction than , and vice versa;—that the sign of a square root has a natural fitness for indicating the taking of a logarithm; and many other things, for all of which it is our firm intention to find good reasons. We mean to avow broadly, that it is as absurd to keep to any standard of notation, as it is to expect uniformity of opinion in matters of

reasoning: and we almost hope to discover that it would facil-
itate communication, if no two writers used any one symbol in
common.

The average standing of the gentlemen who have used or advocated the *entire new notation*, so far as we know them, is under
four years from the time of the M.A. degree. We see therefore
that, as yet, the younger members of one college have it all to
themselves. We state this in praise of their zeal if their system
be an improvement, or in blame of their presumption, if it be no
such thing, and that they may have all the credit in either case. It
will be no small honour to them if they can succeed in changing
the system of the University. We have declared our own opinion,
and we see with pleasure that in an original and excellent work,
on the elements of electricity, &c. from Caius College, published
since the *début* of the new system, it is not adopted. But we again
call upon those who have influence, to use all fair means either
to check or to promote the alteration. To have no community of
system—to have the moderators of one college using one, and
those of another using another, each forcing his own upon the
whole University during his year of office,—to oblige the same
student to read books of different notation, because each happens
to be the best of its kind—to keep him in suspense as to what
notation he will be examined in, till the beginning of his fourth
year, when he ought long before that time to be thoroughly well
grounded in one or other—will be no advantage to the cause of
science in Cambridge.

The first textbook to use the symbol d_x

It would appear that Earnshaw wrote his tract for recommending to his peers the Lagrangian version of calculus and the new
symbol d_x; this is probably why he did not find it necessary to
show that Lagrange's definition of the differential coefficient (or
derivative) was compatible with that in which the derivative was
taken as the limiting value of the incremental quotient.

So far as I can tell, the first elementary treatise where d_x served
as the symbol for the differential quotient was written by W. H.
Miller (after whom the *Miller indices* are named) and appeared in

1833 [163]. Eq. (β) in §2 of Miller's book, reproduced below,

$$\therefore f(x+h) = f(x) + f'(x)h + f''(x)h^b + \ldots \qquad (\beta)$$

is the analog of Eq. (12.24c). Miller concludes §2 with the remark: "It will be seen hereafter that cases may occur in which the expansion of $f(x+h)$, in the form given above, becomes impossible when x has a particular value."

The contents of §§3–4 are reproduced below:

3. The coefficient of h in the expansion of $f(x+h)$ is called "the differential coefficient of $f(x)$ with respect to x," and is denoted by the symbol $d_x f(x)$. The differential coefficient of $d_x f(x)$, or the coefficient of h in the expansion of $d_x f(x+h)$, is called the second differential coefficient of $f(x)$, and is denoted by the symbol $d_x^2 f(x)$, and so on.

Using this notation, equation (β) may be written thus

$$f(x+h) = f(x) + d_x f(x)h + f''(x)h^b + \ldots$$

The quantity x, which $f(x)$ contains, or depends upon, is termed the independent variable.

4. $d_x f(x)\{[f(x+h) - f(x)] \div h\}_{h=0}$,

$d_x f(x)\{[f(x+h) - f(x)] \div h\}_{h=0}$ denoting the value of $d_x f(x)\{[f(x+h) - f(x)] \div h\}_{h=0}$ when h is indefinitely diminished.

$$f(x+h) = f(x) + f'(x)h + f''(x)h^b + \ldots \qquad (\beta)$$
$$\therefore d_x f(x) + f''(x)h^{b-1} = [f(x+h) - f(x)] \div h :$$

and since $b - 1, \ldots$ are all positive, $h^{b-1}, \ldots$ vanish when h vanishes, therefore

$$d_x f(x) = \{[f(x+h) - f(x)] \div h\}_{h=0}.$$

Who was Thomas Jarrett?

Thomas Jarrett was educated at St. Catharine's College, Cambridge, and secured a first class in the Mathematical Tripos (thirty-fourth wrangler) and in the Classical Tripos (seventh place) in 1827 [164, pp. 486, 603], the year in which De Morgan was the fourth wrangler. The following year he was elected a Fellow of his college, and became Sir Thomas Adams's Professor of Arabic in 1831 and Regius Professor of Hebrew in 1854. Jarrett made an extensive study of algebraic notation, which he discussed first in an article [160], and at greater length in a monograph [161]. He was known more as a polyglot interested in the translation and Romanization of foreign languages than as a mathematician.

In the first work, some symbols of our interest were explained as follows [160, p. 74]:

27. The letter D being considered as an abbreviation of the word *difference*, the symbol $D_x^n \cdot u$ will denote the n-th difference of u, x being the independent variable and the increment of x being arbitrary. The case in which Dx, or the difference of x, is unity being of very common occurrence, we may denote by $\Delta x^n \cdot u$, the n-th difference taken on that supposition.

29. By a slight modification of a notation proposed by Euler we may greatly abridge the method of expressing partial differential coefficients. Instead of $\dfrac{d^{m+n}u}{d^m x \cdot d^n y}$ he proposed to write $\dfrac{d^m}{x} \cdot \dfrac{d^n}{y} \cdot u$; if we change this last into $d_x^m d_y^n \cdot u$ we get an expression much more simple. Should the quantity to be differentiated be a power of the independent variable itself, the index subscript may be omitted; thus $d \cdot x^n$ may be written for $d_x.x^n$.

Note to Art. 29 [p. 103]. I was not aware till a few weeks since that this notation is not original. It is just hinted at by Lacroix, *Traité du Cal Diff. et Int.* Tom. II. p. 527.

In the second work, he added three explanatory notes for drawing attention to those who had already anticipated three of his symbols, albeit fleetingly; the text of the second note is reproduced below [161, p. iv]:

(2) $d_x^n u$, for $\dfrac{d^n u}{d x^n}$, is due to Lacroix, although not used by him, being merely pointed out in a single line [footnote: Calcul Diff. Tome II, page 527.]; it was suggested to the writer of these pages by the analogous integral notation invented by Professor Airy.

Chapter 13

Airy's Allies and Adversaries

G EORGE Biddell Airy introduced his symbol for integration in 1826 in a paper on the figure of the earth [165] and, as already mentioned, in the first edition of his *Tracts* [56]. Two passages from the latter source have already been quoted; we will now see how he introduced the new symbol in the paper:

…that is, we must take $\int_a \int_\omega' \int_\mu' \frac{\rho R^2}{z} \cdot \frac{dR}{da}$ *, or in the common notation $\iiiint \frac{\rho R^2}{z} \cdot \frac{dR}{da} \cdot da \cdot d\omega' \cdot d\mu'$. The limits of μ' are -1 and $+1$: those of those of ω' are 0 and 2π: and those of a are 0, and the value of a at the external surface, which we shall call a. [*Footnote: I prefer this notation, as it does not necessarily carry with it the infinitely small quantities.]

The word "allies" has been put in the title more for strengthening the alliteration than for the sake of suggesting that Airy had a large number of allies. No purpose will be served by naming *all* those who adopted the subscripted integral sign, even if they were known to me. Let us begin by seeing what his critics wrote.

207

W. R. Hamilton's letter to P. G. Tait [166, p. 121]

Your *name* was familiar to me, before Dr Andrews was so good as to propose that we should have some personal acquaintance with each other. But I regret (and perhaps ought to be ashamed) to say, that as yet I have not had an opportunity of reading any of your *work*. However from the specimen sheet which you sent me, along with your first letter, of a book of yours on analytical mechanics, & in which you did me the honour to introduce the subject of the Hodograph, I *collect* that you consider it *judicious, at least* (if not absolutely necessary) *in instruction*, to use *differential coefficients only* & to exclude differentials themselves. And perhaps you may have *adopted*, even publicly as Airy has done, using the (to me) uncouth notation $\int_\theta f(\ )$ for $\int (\ )\, d\theta$ the *system* which *rejects differentials*. If so, I can only plead that I am not intentionally, nor knowingly, *controverting* anything which you have published. And if I now quote Moigno, it is merely to show that *I am not wishing to be singular*.

Augustus De Morgan [89, pp. 198–9]

We have reserved the notion of differentials (which we may abbreviate into diff[ls].) as distinguished from diff. co. [meaning *differential coefficient*], till we have come to a point at which the occasional rejection of the latter in favour of the former will produce an advantage more than compensating the liability to inaccuracy which the former are said to involve.[1] (Read here pages

---

[1]The author takes this opportunity, once for all, to dissent from notions which have been lately promulgated in English works, relative to the rejection of differentials. To such a point has this been carried, that the very striking and instructive analogy between $\Sigma y \Delta x$ and $\int y\, dx$, as compared with that which exists between $\frac{\Delta y}{\Delta x}$ and $\frac{dy}{dx}$, has been lost to the eye by the introduction of $\int_x y$ to stand for $\int y\, dx$. But has this great sensibility to notation been accompanied by a similar feeling with regard to the assumption of principles or theorems? Have those who imagined they were more accurate when they wrote $\frac{dy}{dx} = p$ instead of $dy = p\, dx$, rejected the assumption that $f(x+h)$ can always generally be expanded in whole powers of h, or the attempts at *a priori* proof, after the manner of Lagrange, that fractional and negative powers cannot enter? And have they been equally attentive to phraseology? Have they rejected

14, 15, 38 41 of the Elementary Illustrations.) If $u = \phi(x,y)$ give $\Delta u = \phi' + \phi, \Delta y + \&c.$ (page 87) we write $du = \phi' \, dx + \phi, dy$ as an equation 1, which approximates without limit to truth, as dx and dy are diminished; 2. as one which gives the limits, so soon as ratios are formed by division, upon all suppositions. The only warning necessary is, never to separate a partial differential from its denominator without making a proper distinction, since the removal of the denominator removes the existing distinction. Thus $du = \dfrac{du}{dx} dx + \dfrac{du}{dy} dy$ cannot be written $du = du + du$, though we have du(when both vary) $= du$ (when x only varies) $+du$ (when y only varies).

Which might be written $du = d_x u + d_y u$, but $du = \dfrac{du}{dx} dx + \dfrac{du}{dy} dy$ will be found more convenient.

J. J.'s *Observations on the notations employed in the differential and integral calculus* [167]

The differential and integral calculus are applied to nearly the whole circle of the physical sciences; scarcely any treatise on mechanics, optics, astronomy, &c. can be read so as to be understood without a thorough knowledge of these extensively useful adjuncts, or at all events without a pretty close acquaintance with them. It is clearly expedient then that sciences so generally applied and so constantly occurring should be kept as simple as possible. The symbols employed should be as free as they can be from ambiguity, at the same time, there should be nothing cabalistic or mystifying about them. The sciences are nearly universal in their application, so likewise should be their notation; and to this end, there should be a sort of unity about it which would at once identify it; so that when a reader opens a scientific treatise

---

the expressions about the *failure* of Taylor's Theorem, which would imply that the said expansion, not having the process by which it was declared universal before its eyes, but being moved and instigated by the vanishing of a factor, did wilfully and of malice aforethought, refuse to be true in Chapter II., the same being against the proof in Chapter I., its truth and generality? Until these questions can be answered in the affirmative, we are reminded of differentials by the relative sizes of a gnat and a camel [89, p. 198].

he may know at a glance what calculus is adopted in its demonstrations. He can then begin to read it, but obviously this cannot
be done, if he have been accustomed to one kind of notation and
a totally different one be used in the book: he must in the first
place learn his letters, and *if no explanation be given*, it may require much time and trouble to bring him acquainted with an
old friend disguised in a new dress: he may have learned Greek
and be competent to read that language; but he may not be able
to read the same thing in Hebrew characters. Clearly, if one kind
of symbolical language expresses either of the sciences named
more accurately or *more logically* than another, that language ought
to be generally adopted, and no other used: such language ought
to make its appearance in every treatise having any pretensions
to elegance, and all others be made over for the exclusive employment of scientific charlatans.

Could writers on the differential and integral calculus agree
upon the point, as to which is the most accurate mode of expressing the various processes to which they are applied, and *use no
other*, and would the authors of other scientific works adopt only
the language thus set apart, they would very much indeed simplify those important sciences, as well as their applications: they
would save their young readers a great deal of useless trouble:
they would also, by giving a oneness and a generality to the
symbols employed, remove from those sciences that *shifting*, or
as some term it *hocus pocus* [sic] sort of character, to which their
appearing now in one form and then in another certainly entitles
them.

Moreover, if one has learned to read a mathematical process
in one symbolical language, it would be difficult to prove how it
adds a particle to his knowledge to be able to read it in another;
and therefore the time and the trouble that it costs him in learning
to read the process in its new dress is time lost and labour thrown
away. It is supposed that this position will not be disputed by
the advocates of either notation, and if this be the case, it surely
behoves men of competent authority to consider the subject with
the view of rescuing it from such a stigma; it is hoped that they
will endeavour to prevent the votaries of science from having
their time thus uselessly wasted; from being needlessly puzzled
by different notations or bewildered by a mixture: to realize this

hope, by calling due attention to the matter, is the object of the preceding and subsequent remarks.

Up to a recent period the fluxional notation was commonly used by English mathematicians. Mr. Woodhouse assigned reasons for the adoption of the differential instead of the fluxional notation in the preface to his Principles of Analytical Calculation, published in 1803; he employed the differential method in an elaborate paper published in the Philosophical Transactions in the next year: previously he had used the fluxional notation. The English translation of Lacroix was published in 1816; the differential notation first occurred in the Cambridge Problems in 1817. I believe its first appearance in any English mathematical periodical was in the second volume of the Mathematical Repository, in a solution by Mr. Ivory.

In the translation of Lacroix's Differential and Integral Calculus just named, it was laid down that if u be a function of x and $u = ax^3$, then $du = 3x^2dx$, and $\frac{du}{dx} = 3ax^2$; the first expression was termed the "*differential*" of the equation, the latter was called its "*differential coefficient*."

I believe this notation has generally been since used by writers on the differential calculus, both in England and elsewhere; another mode of differentiating, however, has been partially adopted at Cambridge, or perhaps it may be more accurately termed a substitute for differentiating; it has been called "*the calculus of differential coefficients*:" instead of writing $\frac{du}{dx}$, for the differential coefficient as above, they write $d_x u$: if u and z be functions of x, they write

$$d_x(uz) = u d_x z + z d_x u.$$

Similarly, $d_x(u^z) = u^z \left(z \frac{d_x u}{u} + \log_\varepsilon u \, d_x z \right).$

The radius of curvature is thus expressed:

$$r = -\frac{1}{d_{xy}^2} \left[1 + (d_x y)^2 \right]^{3/2}.$$

The equations of motion are thus written:

$$d_t x = \text{velocity parallel to } x. \tag{13.1}$$
$$d_t y = \text{velocity parallel to } y. \tag{13.2}$$
$$d_t z = \text{velocity parallel to } z. \tag{13.3}$$
$$d_t^2 x = X \, d_t^2 y = Y \, d_t^2 z = Z. \tag{13.4}$$

It is well known that these expressions are usually written,

$$\frac{dx}{dt}, \quad \frac{d^2 x}{dt^2}, \quad \&c.$$

Perhaps Mr. Jarrett's paper on algebraic notation in the third volume of the Cambridge Transactions, printed in 1827, contained the first specimen of the index subscript-notation, though he says Prony and others had previously employed it; the subscript notation, or the calculus of differential coefficients, has found its way into some treatises on mathematical subjects. It is supposed the above examples will be sufficient to indicate the difference between the subscript notation and that more generally used; but in order to become a proper judge of the difficulty in reading a book written in the new notation, after one has been accustomed to the common one, the reader must go through the task himself; and his qualification to give an opinion will be all the better, if he have to commence with finding out what the new notation really means.

For Jarrett, see p. 205ff.

The notation of Lacroix (that is the notation employed by him) has been so generally used in mathematical works during many years, that some strong reason ought to be given for introducing another; on this head, however, I have met with only one advocate, namely, the author of a small work entitled "On the Notation of the Differential Calculus." The work is said to be scarce; my copy has no title-page, but the book was printed by Metcalf, Cambridge, some time ago. The author is understood to be a very distinguished member of the University; the reader should refer to the book for practical illustrations and for the full scope of the writer's object; only some extracts, strictly bearing upon the point under consideration, can be taken on the present occasion.

Earnshaw's pamphlet

Art. 32. The author says, "We must observe that since du

is obtained from u by performing upon it some operation with regard to x, of which it is a function, it is necessary when u is a function of several independent variables $x, y, z, \ldots$ to know with regard to which of them the operation d is to be performed, for the results may be very different according as d is performed with regard to the one or the other. And herein the notation of differentials is defective, for when we meet with the expression du it is impossible to know what it represents. We know, indeed, that an operation of a certain kind is to be performed on ic in respect to some quantity of which u is a function, but which is that quantity it is impossible to tell. Hence arises all that confusion and obscurity from which very few, if any, treatises on differentials are free; and in this respect also it is very much inferior to the calculus of differential coefficients, which is remarkable for its perspicuity."

Art. 42. "Whenever u is a function of one independent variable, the differential coefficients may be represented by the fractions $\frac{du}{dx}, \frac{d^2u}{dx^2}, \frac{d^3u}{dx^3}, \ldots$

"But if u be a function of two independent variables x and y, then, because $du = d_x u\, dx + d_y u\, dy$, we have

$$d_x u = \frac{du}{dx} - d_y u \frac{dy}{dx}$$

from which it appears that the differential coefficient $d_x u$ can no longer be represented by the fraction $\frac{du}{dx}$. Hence the only notation which is inconsistent with itself, and consequently erroneous in principle, is precisely that which is most used in this University, viz. the representing differential coefficients by $\frac{du}{dx}, \frac{d^2u}{dx^2}, \ldots$

"This ought to be a sufficient reason for rejecting it and endeavouring to invent another which shall at least be consistent with itself. * * * * * * * * *

"The ridiculous subterfuges to which writers have been driven by the use of $\frac{du}{dx}, \frac{d^2u}{dx^2}, \ldots$ render it a matter of wonder that that notation has not long ago been banished from every mathematical treatise."

The writer concludes his work by giving the following examples of the confusion arising from the common, which he terms *inconsistent notation*.

"At page 175 of the Cambridge translation of Lacroix's Differential and Integral Calculus, we have these two equations,

$$\frac{du}{dx} + \frac{du}{dz} \cdot \frac{dz}{dx} = 0, \text{ and } \frac{du}{dy} + \frac{du}{dz} \cdot \frac{dz}{dy} = 0$$

on which the author remarks, 'the dz of the first equation must not be *confounded* with that of the second.' Now we ask what is there to distinguish dz in the one from dz in the other? Nothing. In fact, this remark alone ought to have been sufficient to demonstrate the necessity of an improvement in the notation. A little below, in the same page, we find the two following *explanatory* equations,

$$dz = \frac{dz}{dx} \cdot dx \text{ and } \frac{dz}{dy} \cdot dy$$

which we hold to be utterly *unintelligible*, though they are given by way of *explaining* the mystery of their predecessors above.

"We shall take our next example from a book, the title of which it is not necessary to mention. "'z being a function of x and y, two independent quantities, the following equation is said to express the *connexion* between $\frac{dz}{dx}$ and $\frac{dz}{dy}$,

$$\frac{dz}{dx} = \frac{dz}{dx} + \frac{dz}{dy} \cdot \frac{dy}{dx}$$

from which we should naturally conclude that

$$0 = \frac{dz}{dy} \cdot \frac{dy}{dx}.$$

"Now we ask, what connexion does this establish between $\frac{dz}{dx}$ and $\frac{dz}{dy}$. Certainly none. In order, however, to explain how equation (1.) does represent such a connexion, we are told that '$\frac{dz}{dx}$ on the *left*-hand side of the equation does not mean the same as

$\dfrac{dz}{dx}$ on the *right*-hand side;' an explanation not likely to be very satisfactory to a learner."

The above are the strongest reasons, indeed the only reasons, that I have seen advanced for adopting the new notation. I have made these extracts, in order to set the writer's most cogent arguments before the reader; still, I would advise him to peruse the book and form his own judgement.

With regard to the above quotations, I wish briefly to remark, that it seems to me the reference to Lacroix is not sufficiently explicit to do justice to that work. It should be observed that the two differential equations taken from that work belong to two sections of the *curve surface*, the equation of which is $u = 0$; and "the dz of the first equation must not be confounded with that of the second, *for they are both only partial differentials, as has been remarked in No. 120.*"

I do not pretend to determine the point, but I am impressed with the notion that a careful perusal of Arts. 120 and 127 in Lacroix, upon which the equations cited depend, will clear up the mystic appearance which they bear in the pamphlet. With respect to the equations $dz = \dfrac{dz}{dx} \cdot dx$ and $dz = \dfrac{dz}{dy} \cdot dy$, said to be given in Lacroix by way of *explaining* the mystery, this is what Lacroix really does say:—

"When we have $dz = p\,dx$, dz is the differential of the ordinate of the section parallel to the plane of x and z: and similarly, $dz = q\,dy$, dz is that of the ordinate of the section parallel to the plane of y and z;" here $p = \dfrac{dz}{dx}$ and $q = \dfrac{dz}{dy}$: I am unable to perceive that these equations were *intended* to clear up the mystery ascribed to the equations first mentioned. Similar equations to the sections parallel to the planes of zx and zy have been given in Higman's Syllabus, and in other works: perhaps the geometric signification of the equations cited ought not to be overlooked in criticising them. Whether the author, the title of whose book is not mentioned, *himself* clears up the apparent paradox that "$\dfrac{dz}{dx}$ on the *left*-hand side of the equation does not mean the same as $\dfrac{dz}{dx}$ on the *right-hand* side," does not appear. I do not know that I

216

have his book, and therefore shall leave the author to take care of himself.

Perhaps it is impossible fully to illustrate the subject entirely free from paradox: thus the writer of the pamphlet on page 24, says, "the reader must keep in mind, that though x appears in the expression dx, yet dx is entirely independent of x," &c. : if dx have nothing to do with x, the quantity, or whatever x denotes, must have been altogether annihilated, or completely changed, in becoming dx, and in that case x in the latter expression ought to have been some other symbol, to prevent what Berkeley calls a *fallacia suppositionis*, or, "*a shifting of the hypothesis.*"

The writer of the pamphlet in Art. 26, and Professor Miller, See pp. 204–205 for §3 of Miller's textbook (Differential Calculus, Art. 3), state the operations which d_x denotes when affixed to a function of x; there is, or at least I fancy there is, a material disparity in their statements; the reader can if he please turn to the works and judge for himself; my object for naming the circumstance is, because one of the books was written apparently to recommend the new notation, and the other is the only elementary treatise that I know in which it is used.

Having cited the above reasons in favour of the d_x notation, I will now quote two opinions on the other side of the question. Mr. W. S. B. Woolhouse, in a very valuable disquisition on the fundamental principles of the Differential and Integral Calculus, published in the Appendices to the Gentleman's Diary for 1835 and 1836, says, "Before closing this paper I cannot refrain from adding a remark on a new plan of differential notation that has lately been introduced, and to a considerable extent adopted at Cambridge. I allude to the substitution of $d_x y$ for $\dfrac{dy}{dx}$, $d_x^2 y$ for $\dfrac{d^2y}{dx^2}$ and others of a similar kind, which possess no recommendation whatever except it be that of novelty: and I feel convinced that this change is suited only to such persons as are satisfied with mere *hocus pocus* operations on optical symbols without any regard to the mental images they are designed to represent. In the higher branches of analysis this new-fangled notation will defy the presence of anything like distinct ideas: for instance, an elemental parallelopiped $dx\,dy\,dz$ is reduced to the confused and incomprehensible form $d_x y\, d_x z\, dx^3$. It is the duty of every mathematician to make known his opinion concerning it."

Mr. Woolhouse's attainments are such, that his opinion upon this, or upon any other mathematical subject, is certainly deserving of respect.

A writer in the Northumbrian Mirror, new series, p. 89, says, "We cannot conclude without noticing the clumsy differential notation which has recently captivated the publishers of mathematical works at Cambridge; it offends against simplicity, symmetry and clearness; it is a meretricious show of conciseness, and an innovation that every lover of simplicity, brevity and neatness, should repudiate.

"In the differential coefficients of two or more varieties the expression is, of necessity, so overloaded with those little ugly off-shoots growing out of the side and stem that the body of the tree is almost hidden from the view."

I have made this quotation to show the writer's opinion, I certainly do not see that a little x deserves to be called ugly any more than a great one, but the phraseology is the writer's; however, it will be observed that the opinions on each side are made pretty strong.

The integral calculus is the reverse of the differential; integration is commonly denoted by the symbol $\int$; thus,

$$d \cdot \sqrt{a^2 + x^2} = \frac{x\,dx}{\sqrt{a^2 + x^2}} \text{ and } \int \frac{x\,dx}{\sqrt{a^2 + x^2}} = \sqrt{a^2 + x^2}.$$

In the suffix notation for integrals the sign of integration is $\int_x$; in this case only the differential coefficients are employed: if the d_x notation be employed at the same time, then $\int_x$ is the symbol of an operation precisely the reverse of d_x; thus—

$$d_x \cdot \sqrt{a^2 + x^2} = \frac{x}{\sqrt{a^2 + x^2}}$$

and
$$\int_x \frac{x}{\sqrt{a^2 + x^2}} = \sqrt{a^2 + x^2}.$$

The suffix notation for integrals is found in many of the mathematical works at Cambridge; for instance, Hymer's Integral Calculus and Differential Equations, Murphy on Electricity,[2] Earnshaw's Works, Airy's Tracts, &c.

[2]For a detailed citation, see Ref. [168]

The authors just mentioned are an honour to one of the first universities in the world; they are stars of the first magnitude in that firmament of science. Every one must feel that a notation employed by such authors must have some peculiar advantages to recommend it; still I believe neither of these celebrated writers has given any reason for adopting the suffix notation in preference to the common one. I wish they had, but as the matter stands I can only adduce their names in support of the $\int_x$ mode. On the other hand I wish to cite the reasons of two justly esteemed authors for employing the old notation.

Mr. Pratt, in the preface to his 'Principles of Mathematical Philosophy,' ed. 1835, says, "The prevailing argument with me for using the old differential and integral notation is the excellence of Fourier's notation for definite integrals. I much prefer that to any other that I have seen, and this naturally led me back to the old form of differentials and integrals. In case any of my readers are not acquainted with Fourier's notation, I now give it:
$\int_a^b u\,dx$ represents the integral of the differential coefficient u, or of the differential $u\,dx$ with respect to x taken between the limiting values a and b of x. In successive integrations the order of arrangement of the integrals is the same as that of the differentials: thus $\int_{-1}^1 \int_0^{2\pi} P\,d\mu\,d\omega$ represents the double integral of P with respect to x and ω, the limits of x being 1 and 1, and the limits of ω being 0 and 2π."

Mr. Gregory, in the preface to his 'Examples of the Differential and Integral Calculus,' thus expresses himself:— "I have adhered throughout to the notation of Leibnitz in preference to that which has been of late revived and partially adopted in this University. Of the differential notation I need say nothing here, as it appears to be abandoned as an *exclusive* system by those who introduced it; but as the suffix notation for integrals has been sanctioned by those whose names are of high authority, I may state briefly some of my reasons for differing from them.

"In the first place, on considering the subject, I could find no arguments against the use of the notation for differentials, which did not apply with even greater force against that for integrals; indeed, although there may be some cases in which the use of the former is advantageous, I know of none in which the latter does not appear to me to be inconvenient. "In the next place, I fully agree with Professor De Morgan in an unwillingness to lose sight of the analogy to summation which is implied in the old notation : and if it were at any time necessary to consider integration merely as the inverse of differentiation, I should prefer such a symbol as d_x^{-1} which expresses the required idea better than $\int_x$. But what I look on as a fatal objection to the suffix integral notation is, that, like the corresponding one for differentials, it is not applicable to all cases. Of this any one may satisfy himself by attempting to use it in transforming a multiple integral from one system of independent variables to another, a problem which is of frequent occurrence, but which I have not seen solved analytically in any work in which the suffix notation is employed; so long, therefore, as the old notation adapts itself to all cases in which it is required, while that which is proposed is not so accommodating, there appears to me no doubt which is to be preferred." objections raised by Gregory

Mr. Jarrett, in the paper before referred to, recommended Fourier's notation; everyone who habituates himself to its use will, I think, admit that it is admirably adapted to effect its purpose.

In the commencement of these remarks I gave it as my opinion that new symbolic language in a scientific treatise uselessly puzzles the reader and occupies his time without increasing his knowledge; it has been, and will be, my object to make converts to this notion. Some of the writers abovenamed adopt the suffix notation both in differentials and integrals, others only the suffix for integrals: must not even this *variety* be an incumbrance to a student, and for that reason ought it not to be exploded?

Airy, in the commencement of his tracts, says, "By the notation $\int_\theta \cos n\theta \cdot \Theta, \int_\theta \cos n\theta \cdot \cos \overline{m\theta + D}$, &c. we mean what are usually written

$$\int \cos n\theta\, \Theta\, d\theta, \int \cos n\theta \cdot \cos \overline{m\theta + D}\, d\theta, \&c.$$

they are the quantities whose differential coefficients, with respect to θ, are $\cos n\theta \cdot \Theta, \cos n\theta \cdot \cos \overline{m\theta + D}$, &c."

To a reader who is unaccustomed to the notation, and who may perhaps here meet with it for the first time, this explanation is exceedingly useful as a key, it sets him a-going.

In some other books, however, in which the d_x and $\int_x$ notations are used, not one word by way of definition is said: the reader has to find out the meaning of a new symbolic language used in investigating intricate subjects as he best may: it seems to be taken for granted, that because the writer knows the meaning of the hieroglyphics, the reader must comprehend them by sheer intuition.

At college, where lectures are constantly given, such difficulties may be well explained *viva voce*. But the writers of mathematical works at Cambridge should bear in mind that they advertise their books for sale; the celebrity that so justly attaches to the University and its members induces lovers of science who are at a distance from the University, and who are not acquainted with its advantages, to purchase their works; they cannot attend the lecture-room, nor is a private tutor at hand; the explanation, therefore, that a Cambridge man can obtain on the spot should form part of the work; without it, the book at first appears to be a mathematical enigma, the solution of which wastes the student's time without at all adding to his knowledge. It is considered very prudent and discreet in men who have nothing to say, to say nothing, but this is not the case with the Cambridge authors to whom I allude. They are quite capable of writing to the purpose, and might in few words clearly explain a difficulty; in my opinion their doing so would double the value of their labours, and trebly enhance their utility; besides, it is only a matter of common justice to their *general readers*.

It seems to have become the fashion with many writers to eschew all introductory or explanatory matter: the reader, without preamble or preface, is at once pitched into the midst of a difficult subject entirely in a new dress, and he may make progress if he can. Works thus written under the notion that readers intuitively know everything without being told anything, may have more matter compressed into them, but their authors lock it up and keep the key. To explain myself more clearly, I will give an example: let any one who knows nothing of the suffix notation, but who can read, for instance, a treatise on dynamics in the old, take a treatise on that subject written in the suffix language, containing no explanation whatever; let him move on as fast as he can, and tell me how he likes it.

If the reader has never met with a sudden change from one notation to another, he may find a specimen in the treatise on Mechanics in the *Encyclopædia Metropolitana*: up to Art. 56, the differential and integral calculus are employed; subsequently the *fluxional*: thus within a few lines in writing on precisely the same subject, the symbols and language are completely changed. The author of the treatise referred to is deserving of great respect, but I wish he had taken the trouble to write the article all in the same language. By way of parenthesis it may be observed that the *expense* of these books is considerable, and judging from their authors' celebrity, their purchasers have a right to expect the articles to be of a high order, and not mere scraps indifferently *cooked* and sent into the world scarce half-made up; however, many of the treatises are worthy of their authors, and are a credit to the seat of learning from which they emanate.

It has been before remarked, that a new notation is very likely to arrest the progress of an uninitiated reader. I will mention an instance: I some time ago had the good fortune to fall in with an ingenious mechanic at Penzance, who has been for several years in the habit of answering many and sometimes most of the mathematical questions proposed in the Diaries, &c.; he told me that he had taught himself the fluxional calculus, and that by reading the Diaries and other publications of that description, he had learned to read and to use the differential mode; but having never met with any explanation of the suffix notation, he could not at all understand it, nor had he been able with all his application to read a solution written in that language so as to comprehend the steps in the demonstration; it had baffled all his efforts, and he had given up the attempt to find out its meaning.

I for one sincerely respect the votaries of science, even when education, leisure, fortune, and every concomitant tend to place them in the most favourable position. No one values more highly than I do the labours of professors, &c., whose very calling, the sole business of their life, is science; whose minds are very often kept on the stretch as a mere matter of bread and cheese; but if men like these are entitled to esteem for their mental

222

endowments—and I trust they always will be—is not the man deserving of it who, after his daily manual toil of a laborious description, devotes his hours to science and, notwithstanding every disheartening difficulty, works his way up to her most secluded walks? If this be so, should not men eminent by talent as well as by position be careful to throw no impediments in the rugged paths of such sons of genius.

At all events, I think they should; but it is possible that I may have formed a hasty or a partial judgement upon the subject, for I well know many of the difficulties attending such a course, and I readily admit that I have had the case abovenamed in my mind's eye in writing the above remarks; whether 1 have been skilfully keeping to the point so as to effect my purpose or not, is a question not for me to decide: I am quite sure that each and all of the authors that I have mentioned would much more willingly assist such devoted disciples of science as the very remarkable one to whom I have alluded, and indeed all others, than place the least obstruction in their arduous paths.

Many of the most esteemed mathematical authors of the day are correspondents or readers of this publication: the object of this paper has been to bring to their notice the loss of time which attends their using new notations in scientific treatises without any explanation, and also the uselessness, as I think, of adopting different notations to effect the same purpose. Either the common, the suffix, or some other is the best adapted for general use; which ought to be constituted and considered the standard notation, I want the opinion of your readers to settle: to elicit this, I have adduced the pith of what has been written in favour of and against the suffix mode, so that a judgement may be formed with as little trouble as possible: I have endeavoured to do my part as impartially as I could. 1 have aimed to state the case as fairly as I could in hopes to obtain an impartial judgement from competent authority. Very likely I may have evinced a little leaning, for I do not like the suffix notation, but I trust I have said nothing distasteful to its distinguished advocates: I am willing to pay homage to their splendid achievements in science, but whilst I heartily respect their attainments, I have some misgivings as to their orthodoxy respecting the subject that has been discussed.

It may save time and trouble if I beg to be considered as the voluntary, though feeble advocate of such votaries of science as the one referred to above: as far as I am personally concerned, all writers may go on their own way rejoicing; what they may do, or may not do, will not affect me in the least: on my own account I have no favours to ask. I trust, therefore, in dealing with the subject that my mode of advocacy will be entirely overlooked.

Mr. Gregory says the d_x notation is losing ground: if it be not defunct, I trust it will soon become so: no suffix notation appears in the last Cambridge Problems: this gives some hopes. I think there will be a unanimity in the opinion that the fountain head of mathematical science should be kept clear, that its streams should enlighten as they proceed and not darken and mystify. Those who are placed at the source would in my opinion do well to ponder on the influence they have on the scientific world, to see how much it behoves them to throw a light upon all the subjects they handle, but especially to refrain from making them repulsive by clothing them with darkness.

Should the preceding desultory remarks educe opinions which may settle a point that now appears to be doubtful, or should they have the least tendency to fix the adoption of one notation, and thus to simplify the science named at the commencement, my aims will have been fully accomplished.

Gregory's objections

Two objections, raised by Gregory in the preface to his *Examples* [62], were mentioned by J. J. (see pp. 218–219 above). I am in agreement with the first (that if d_x is used as the symbol for the derivative, then d_x^{-1} expresses the idea of an antiderivative better than $\int_x$). Recall, however, that De Morgan objected to the use of d_x itself, and argued that D_x would have been an acceptable choice. Since D_x and D_x^{-1} have now gained currency, both Gregory and De Morgan would be satisfied with these symbols. Gregory's other objection—that Airy's symbol cannot handle co-ordinate transformation in a multiple integral—has been shown in Chapter 5 to be invalid, and the following equations have been derived:

$$\left| \int_{\mathcal{J}y} \int_{\mathcal{J}x} f(x,y) \right| = \int_v \int_u f_2(u,v) \left| \left(\frac{\partial x}{\partial u}\right)_y \left(\frac{\partial y}{\partial v}\right)_u \right| \tag{13.5}$$

$$= \int_v \int_u f_2(u,v) \left| \left(\frac{\partial x}{\partial v}\right)_y \left(\frac{\partial y}{\partial u}\right)_v \right| \tag{13.6}$$

$$= \int_v \int_u f_2(u,v) \left| \left(\frac{\partial y}{\partial u}\right)_x \left(\frac{\partial x}{\partial v}\right)_u \right| \tag{13.7}$$

$$= \int_v \int_u f_2(u,v) \left| \left(\frac{\partial y}{\partial v}\right)_x \left(\frac{\partial x}{\partial u}\right)_v \right| \tag{13.8}$$

where $\qquad f_2(u,v) = f(x,y)$.

It has also been stressed in Chapter 5 that the above set should only be used when one is transforming definite integrals.

American users of Airy's notation

We must begin with **Benjamin Peirce** (1809–80), simply because, in the words of one qualified to judge [169, p. 8], "Mathematical research in American Universities began with Benjamin Peirce". In the 1841 edition of the first volume of *An Elementary Treatise on Curves, Functions, and Forces* [170], he used the symbol d_c for the differential coefficient, but this was changed to D in the 'New Edition' of 1852 [171]. The topic of integration was treated in the second volume [172], which appeared in 1846.

> 1. The *integral of a given differential* is the function of which it is the differential; and *the integral of a given finite function* is the function of which it is the differential coefficient. To integrate is to find the integral. The sign of integration is $\int.$; thus
>
> $$\int . d . x = x, \qquad\qquad \int . d . f . x = f . x ;$$
>
> $$\int . d_c . x = x, \qquad\qquad \int . d_c . f . x = f . x ; \tag{245}$$
>
> $$\int^2 . d_c^2 . x = x, \qquad\qquad \int^n . d_c^n . f . x = f . x, \text{ \&c.} \tag{246}$$

For the reader's convenience, Eq. (245) is expressed below in the current notation:

$$\int Dx = x, \qquad\qquad \int Df(x) = f(x); \qquad (245^*)$$

Peirce always integrated finite functions, and a typical formula in his book is shown below:

$$\int . ax^n = \frac{ax^{n+1}}{n+1}. \qquad (279)$$

Peirce's symbols $d_c.$ (or $D.$) and $\int .$ become adequate when the variable of differentiation/integration cannot be inferred immediately; clearly, each symbol needs an additional clue when, for example, a change of variable is to be made.

In the opinion of more than one competent judge [173, 174], **William Ferrel** (1817–1891), a self-effacing autodidact, did more than any other single person to advance the atmospheric physics, and to place meteorology on firm theoretical foundations. He must have become acquainted with Airy's subscripted integral symbol after he "found in a book store a volume containing Airy's *Figure of the Earth* and *Tides and Waves*" [173]. In an article on the motion of fluids and solids relative to the earth's surface [175], published in the first volume of the *Mathematical Monthly*, he used a subscripted D as the symbol for partial differentiation and Airy's symbol (after adapting it to definite integrals involving three variables) without any explanatory text concerning the meaning of either symbol. Ferrel's notation for definite integrals can be illustrated by displaying the LATEX output for one of his equations:

$$v' = \frac{3R^2}{2m} \int_{0}^{l} {}_N \int_{0}^{2\pi} {}_\phi \int_{0}^{\pi} {}_\theta kv \sin^3 \theta$$

In the same volume, the editor of the journal (J. D. Runkle, whose name appears only in the index) published a piece entitled 'Note on Derivatives' in which he explained the meaning of the symbol D_x, and the advantages of using it, but neither he nor someone else contributed a sequel dealing with integrals/antiderivatives.

William Elwood Byerly (1849–1935), a student and ardent admirer of Benjamin Peirce, was known mostly for his undisguised love of teaching [176], which can be experienced even by reading the two calculus tracts mentioned below. His *Elements of the Differential Calculus* was "intended for a text-book, and not for an exhaustive treatise" [64, p. iii]. Using the symbols D_x and $\int_x$, Byerly covers, in fewer than 300 pages, not only what the title promises, but also a good deal of integral calculus. An updated version of this volume could easily outrival many modern textbooks. In the Preface, Byerly listed some notable features of his book:

> Its peculiarities are the rigorous use of the Doctrine of Limits as a foundation of the subject, and as preliminary to the adoption of the more direct and practically convenient infinitesimal notation and nomenclature ; *the early introduction of a few simple formulas and methods for integrating*; a rather elaborate treatment of the use of infinitesimals in pure geometry; and the attempt to excite and keep up the interest of the student by bringing in throughout the whole book, and not merely at the end, numerous applications to practical problems in geometry and mechanics. (My italics.)

Immediately after introducing differentials, Byerly informs the reader that the derivative will thenceforward play second fiddle [64, Chapter 11]:

> 180. The differential notation has the advantage over the derivative notation, that it is apparently simpler, and that the formulas in which it is used are more symmetrical than those in which the other notation is employed; and although the differential is defined by the aid of the derivative, and the formulas for the differentials of functions are obtained from the formulas for the derivatives of the same functions, *there is a practical advantage*, after the formulas have once been obtained, *in regarding the differential as the main thing, and looking at the derivative as the quotient of two differentials.*

A companion volume entitled *Elements of the Integral Calculus* appeared in 1881 [65], followed by a second edition in 1888 [66]. After the first two chapters, which are of a preliminary nature, the third chapter, titled "General Methods of Integrating", begins with Article 49. In quoting the text of this article below, the notation will be modernized; this means that all dots between various symbols will be dropped and $f.x$ will be replaced by $f(x)$;

the reader should note that Byerly refers to the earlier book on differential calculus as I.

49. We have defined the *integral* of any function of a single variable as *the function which has the given function for its derivative* (I. Art. 53); we have defined a *definite integral* as the limit of the sum of a set of differentials; and we have shown that a definite integral is the *difference between two values of an ordinary integral* (I. Art. 183).

Now that we have adopted the differential notation in place of the derivative notation, it is better to regard an integral as the inverse of a *differential* instead of as the inverse of a *derivative*. Hence the integral of $f(x)\,dx$ will be the function whose differential is $f(x)\,dx$; and we shall indicate it by $\int f(x)\,dx$. In our old notation we should have indicated precisely the same function by $\int_x f(x)$; for if the derivative of a function is $f(x)$ we know that its differential is $f(x)\,dx$.

Chapter 14

John West's Posthumous Book

T HE Reverend John West, an impecunious but exceedingly talented and learned Scottish priest, who died in 1817 in Jamaica [177], wrote what appears to have been the first book in English language dealing with the nitty-gritty of *Analysis*, the name applied in his days to Continental calculus [57]. Due to his circumstances, this book could not be published until 1838. He chose to work exclusively with derivatives, and introduced the terms "derivation" (instead of differentiation), "derivative function" (instead of derived function) and "derivative equation" (instead of differential equation). What I have called the "primed function" was named the "prime function" by West, and he applied to the "unprimed function" standard terminology by calling it the "primitive". Considering how fastidious he was, I find it hard to understand that he retained the term "integration" for the "inverse method of analysis", which he defined in words that are reproduced in a later section (see pp. 232–234).

West's nomenclature

West uses the term *derivation* for *differentiation*; if $f(x)$ is a function of x whose derivative is denoted by $f'(x)$, he calls $f(x)$ the *primitive* and $f'(x)$ the *prime function*. He follows Lagrange by seeking a series expansion for $f(x + h)$ in ascending powers of h, and begins his treatment by applying this strategy to the following functions: x^n, a^x, ln x, sin x and cos x. His own notation for the last three functions—namely log. x, sin. x and cos. x—will not be used when he is quoted. Likewise, we will write $f(x)$, $f(y)$ ...instead of the symbols fx, fy ...used by West. Here is an excerpt from his chapter on "Derivation" [57, pp. 21–23]:

> 34. When x is considered as the *original* variable, of which all others are functions, then the prime function of $f(x)$, any function whatever of x, is denoted simply by $f'(x)$. But, if x is to be considered as itself a function of some other variable, the prime function of $f(x)$, it appears, must, in this case, be expressed by $x' \cdot f'(x)$. [cf. Eq. (14.2) below.]
>
> We may therefore consider x as the function of some other variable, and render the preceding rules for finding the prime function more general, if we only multiply by x' each of the prime functions found by these rules.
>
> 35. Again, let $f(y,z)$ denote any compound function of y and z, (which are themselves functions of x) and its prime function will be expressed by $y' \cdot f'(y) + z' \cdot f'(z)$, where $f'(y)$ is used to denote the prime function of $f(y,z)$, when z is considered as a constant quantity, and $f'(z)$ to denote the same, when y is considered constant.
>
> . . .

In § 36, West gives a long list of derivatives; items number 1, 3, and 6 from the list are shown below:

PRIMITIVE	PRIME FUNCTION
1. x^n	$nx^{n-1} \cdot x'$
3. sin x	$\cos x \cdot x'$
6. xy	$xy' + x'y$

Chain rule with an unspecified independent variable

Let us recall that if $y = f(x)$ and x itself is a function of another variable, which we will represent by the symbol η, then

$$D_\eta f(x) = D_x f \cdot D_\eta x = f'(x) \cdot x'(\eta). \qquad (14.1)$$

As Eq. (14.1) is true for any variable of differentiation (provided that $x(\eta)$ is differentiable), we can drop η and state it as

$$Df(x) = f'(x) \cdot x', \qquad (14.2)$$

which is a paraphrase (in symbols) of how West expresses the chain rule (see §34 of his book reproduced on p. 230). We also note for later use that

$$\frac{Dy}{Dx} = D_x y = \frac{dy}{dx}. \tag{14.3}$$

In general, the derivatives considered by West are expressions obtained by differentiating the given primitive with respect to an unspecified variable, and are expressed in the form $f'(x)x'$, and a primitive is found by integrating the given derivative with respect to the same unspecified variable:

$$\int Df(x) = \int f'(x)x'. \tag{14.4}$$

If one is able to find a function $\Phi(x)$ which is such that $\Phi'(x) = f'(x)$, then x can itself be taken as the integration variable

$$\int f'(x)x' = \Phi(x).$$

Of course, $x' = 1$ in this case, but the presence of x' serves to indicate the variable with respect to which integration is carried out. If $\Phi(x)$ cannot be found easily, but a change of variable from x to z, with $z = \zeta(x)$ and $x = \chi(z)$, is found to lead to the result

$$f'(x)x' = f'[\chi(z)]\frac{x'}{z'}z' = \Psi'(z)z', \text{ say,}$$

one can take z as the integration variable, so that $z' = 1$, and

$$\int f'(x)x' = \int \Psi'(z)z' = \Psi(z)$$

The analogy with the notation of d-Calculus can be made manifest if we replace x' with Dx and z' with Dz, and rewrite

$$\int f'(x)Dx = \int f'[\chi(z)]\frac{Dx}{Dz}Dz = \int f'[\chi(z)]\frac{dx}{dz}Dz.$$

West speaks for himself

The stage is now set to examine how West himself presented integration [57, p. 184]:

1. Having shewn how to find the prime function of a given primitive, and all other derivative functions, we proceed to treat of the inverse method, termed INTEGRATION, viz. that of finding the primitive of a given prime function; or rather, of finding the variable part of the primitive. For, if a constant quantity make part of the primitive, as no vestige of it appears in the derivative function, it must remain undetermined. To find the variable part, however, is sufficient, as the constant part may be ascertained from the nature of the problem, for the solution of which, the complete primitive is required.

The inverse method is more difficult, as well as more extensively useful, than the direct; and can only be pursued, as far as it is practicable, by rules reversing the preceding operations; but, in most cases, recourse must be had to methods of approximation.

The prime function, for example, of y^n, a given power of y, being $ny^{n-1} \cdot y'$, it is easy to deduce this rule for finding the primitive, in similar cases, from the prime function, viz. Add 1 to the index of the variable quantity, and divide the function so changed by the increased index multiplied into the prime function of the root; $\frac{ny^{n-1+1}y'}{n-1+1 \cdot y'} = y^n$. And this rule holds, whether the index be affirmative or negative, integral or fractional. There is only one case in which it fails; and that is, when the index of the variable is -1: for, in this case, the primitive of $y^{-1} \cdot y'$, $\left(\text{or } \frac{y'}{y} \right)$ becomes, according to the rule, $\left(\frac{1}{0}, \text{ or} \right)$ infinite. We should therefore have been led to conclude that $\frac{y'}{y}$ had no finite primitive, had not the preceding investigation discovered to us its primitive (log. y) among a different kind of functions.

We shall now set down some formulae, to serve in place of rules, of three different kinds of functions, distinguished by the terms Algebraic, Logarithmic and Circular; and by a few examples under each, point out the method of transforming (if possible) given prime functions into such as shall be similar to the formulae, so that their primitives may be discovered; referring such cases as are more difficult, or altogether impracticable by this method, to be treated of more fully afterwards, when some preparatory subjects shall have been discussed.

The integral or primitive of any function is marked by this sign $\int$ placed before it; thus, $\int y' = y$. $\int (2yy' + 2zz') = y^2 + z^2$, &c. And it is to be observed, that $\int ay' = a \int y'$, $\int c^2 xy' = c^2 \int xy'$, &c.

2.   I. ALGEBRAIC FUNCTIONS.

1. $\int (ax' \pm by') = ax \pm by.$ 3. $\int (xy' \pm yx') = xy$ (14.5)

2. $\int ax^n x' = \dfrac{a}{n+1} x^{n+1}.$ 4. $\int \left(\dfrac{x'}{y} - \dfrac{xy'}{y^2} \right) = \dfrac{x}{y}$ (14.6)

EXAMPLE 1.

Required the primitive of $-\dfrac{xx'}{\sqrt{a^2 - x^2}}$.

Put $a^2 - x^2 = z^2$; then $-xx' = zz'$, and $-\dfrac{xx'}{\sqrt{a^2 - x^2}} = \dfrac{zz'}{z} = z'$;

therefore $\int -\dfrac{xx'}{\sqrt{a^2 - x^2}} = z = \sqrt{a^2 - x^2}.$

We note once again that West does not specify the variable with respect to which the differentiation or integration is to be performed. This means that a reader who wishes to evaluate the integral

$$\int \cos x \, x'$$

must decide themselves whether $x' = 1$ or $x' = Dx$ ($= D_\eta x$, where η stands for some unspecified variable on which x depends). It is natural to settle for $x' = Dx = 1$ (that is $\eta = x$), and conclude that

$$\int \cos x \cdot x' = \int D \sin x \cdot Dx = \sin x;$$ (14.7)

but the result would not change if one chooses z as the variable of integration, taking z to be at first a function of the unspecified

234

variable η, and then choosing $z = \eta$ by setting $Dz = 1$:

$$\int \cos x \cdot x' = \int D_x \sin x \cdot Dx = \int D_x \sin x \cdot D_z x \cdot Dz \quad (14.8)$$

$$= \int D_z \sin x \cdot Dz = \int D \sin x = \sin x \quad (14.9)$$

Who remembers John West now?

Hardly anyone. Since a detailed and balanced assessment of West's achievement has already been published [177], no attempt will be made to reassess his place in the history of British mathematics. He was perhaps the most advanced 'analyst' of early nineteenth century Britain, but worked mainly on his own; unable to participate in the rejuvenation of British mathematics, he has not received adequate recognition, not even posthumously.

In the anonymous article on "Taylor, Brook; Taylor's theorem" in *The Penny's Encyclopædia*, one finds the following footnote [178, pp. 129–30]: "There is a great deal on the subject in the 'Mathematical Treatises' (posthumous) of the Rev. John West, published at Edinburgh in 1838. Mr. West has substituted a notation, for that of Arbogast, in which he will probably have few followers. The student who is not repelled by this, and cannot procure Arbogast's work, will find West's treatises abounding in derivations." The author of the article, it seems safe to assert, must have been Augustus De Morgan.

Chapter 15

Menger's Version of Calculus

Many teachers, upon finding that their students meet difficulties, blame the students themselves for all such difficulties, and blithely go on in their accustomed way. Menger, feeling that such a response was inadequate and unjust, analysed all aspects of the situation, and concluded that the traditional way of presenting the calculus was deeply flawed.

Golland, McGuinness and Sklar [179, pp. xvii–xviii]

S HORTLY after the start of the second half of the twentieth century, Karl Menger published a book [28–30], entitled *Calculus: A Modern Approach* (hereafter abbreviated as *CalcAMA*), that was unmistakably different from any other calculus text. Menger's aim was to avert petrification of calculus, which he saw as the greatest danger that may befall a discipline when its practitioners become complacent. In the preface to the second edition (1953) of *CalcAMA*, he wrote: "A situation which rules out any essential progress can easily arise when specialists abstain from a critical study of books on a subject and at best skim the pages with an eye toward their possible use as texts for sophomore classes". Menger believed that the classical notations, though adequate for routine applications of the rules for juggling with symbols, conceal rather than reveal any understanding of the principles behind these manipulations. It would not surprise

236

the reader to learn that a reform of the kind envisaged by Menger could not be undertaken without an overhaul of the existing notation. He stated that questions of notation play a merely subsidiary role and are important only as they aid in an understanding of the underlying concepts, but the mathematical community at large did not take kindly to his innovations.

Though *CalcAMA* received a modest number of detailed, perspicuous and (on balance) laudatory reviews from fellow mathematicians, the book, and in a certain sense even Menger, are largely forgotten. A biographical sketch by Seymour Kass began by drawing attention to this neglect [180]: "Karl Menger died on October 5, 1985, in Chicago. Except in his native Austria, no obituary notice seems to have appeared. This note marks ten years since his passing." According to Kass, Menger sent a copy of his book to Einstein, who replied that he liked it and recognized the need for a clearer notation, but advised the author against excessive "housecleaning".

After stating that Menger was greatly saddened by the cool response accorded to *CalcAMA*, and that the "failure of the calculus endeavor strained his relations with the mathematical community", Kass goes on to describe the personal traits of Menger:

> His office was a showplace of chaos, the desktop covered with a turbulent sea of papers. He knew the exact position of each scrap. On the telephone he could instruct a secretary exactly how to locate what he needed. Once, in his absence, a new secretary undertook to "make order", making little stacks on his desk. Upon his return, discovering the disaster, he nearly wept, because "Now I don't know where anything is."

Of all the persons in this planet, Menger, surely, should have been the last to be saddened or surprised by the stand-offishness of messy mathematicians towards a person who wanted to tidy up their cosy palace!

My estimate of *CalcAMA*, though favourable, falls short of the praise showered by the reviewer who concluded his review with the following remarks [181]: "*Every teacher of the calculus must read it* [emphasis, not in the original, indicates my agreement]. And he will be thick-skinned indeed who, having read it, does not seriously rethink his lectures; the thinner-skinned reader will rewrite his—and the reviewer confidently believes he will rewrite them à la Menger."

Menger's remarks on the notation of H-calculus

Chapter 6 of *CalcAMA* is entitled "The Basic Concepts of Calculus" and the title of last section is "Remarks Concerning the Classical Notations in Calculus". An experienced reader will be able to read this section without going through the preceding pages, and those who have not Menger's book will be amply rewarded acquainting themselves with his remarks on the notations of classical calculus (his term for what I have called H-Calculus). The chapter ends with the following assessment [30]:

> The designation of antiderivatives as indefinite integrals and of derivatives as differential quotients has had two important consequences.
>
> On the one hand, that notation camouflages the fundamental result of calculus and (as will be seen in Chapter VIII) some difficult operations on derivatives as truisms. In past centuries, undoubtedly, this very camouflage secured the basic ideas of calculus acceptance by many who would have shunned a method obviously surpassing their understanding. Without the protection of a plausible appearance, those great ideas might not even have survived just as some bright butterflies would perish if, with folded wings, they did not assume the appearance of inconspicuous leaves.
>
> On the other hand, the traditional notation has made it difficult to understand calculus. The symbol $\int_a^b f(x)\,dx$, while objectionable on account of its dummy part, is at least reminiscent of the sums of products as whose limit the integral is defined. The symbol $\int f(x)\,dx$, however, not only fails to remind one of the inverse of derivation—the operation he has to perform—but strongly suggests sums of products (the same sums of which the integral is reminiscent) with which the antiderivative concept has absolutely nothing to do.
>
> The traditional symbols, introduced essentially by Leibniz, make it hard to distinguish definitions from theorems, technical difficulties from profound problems, and minor results from tremendous discoveries.
>
> All in all, Leibniz' notation accounts for what, from a sociological point of view, are the two most striking facts in the history of calculus: that for centuries the use of that great theory

has been enormously wide, and that even today its use is often merely mechanical.

Variable-free calculus

A formula that would be written, according to the notation followed in the present book, as

$$D_x \sin x = \cos x \quad \text{and} \quad \int_0^{\pi/2} \cos x \, dx = \left[\sin x \right]_0^{\pi/2}$$

is written by Menger as [30, p. xiii, Eq. (4)]

$$\mathbf{D} \sin \mathbf{x} = \cos \mathbf{x} \quad \text{and} \quad \int_0^{\pi/2} \cos \mathbf{x} \, d\mathbf{x} = \left[\sin \mathbf{x} \right]_0^{\pi/2}. \qquad \text{(KM:4)}$$

The difference may be summarized as follows. Menger uses italic font for *all* functions (including the log and trigonometric), and roman font for quantities that he calls numerical variables. Whereas we write "sin a", by which we mean "the value of sine when its argument is the number *a*", Menger writes "*sin* a". We write D_x or D (if the identity of the independent variable is in no doubt) in italic font, which is usually reserved for variables and functions; Menger prefers to use a font that is upright (not slanted) and bold for the derivative operator, and the symbol **D** will be just as acceptable to Menger.

Menger goes on to suggest that x in his Eq. (4) can simply be shed and the formulas stated therein may be replaced by the variable-free version shown below [30, p. xiii, Eq. (4′)]:

$$\mathbf{D} \sin = \cos \quad \text{and} \quad \int_0^{\pi/2} \cos = \left[\sin \right]_0^{\pi/2}. \qquad \text{(KM:4′)}$$

He continues thus:

> In fact, there is a trend toward actually shedding x after the letters *f* and *g*—symbols that stand for any function or any element of a certain class of functions and which, therefore, have been called *function variables*. In several more advanced books on analysis one can read
>
> $$\text{If } \mathbf{D}f = g \quad \text{then} \quad \int_a^b g = \left[f \right]_a^b. \qquad (5)$$
>
> Only the second formula in (5) contains numerical variables

(namely, a and b) and *g* may be replaced by any function that is continuous between a and b.

Why then is x in (4) usually retained, and, even in more advanced books, reintroduced in applications of (5) to specific functions? The reason is one of the great curiosities in the history of modern analysis—comparable only to the lack of a symbol for the number zero in ancient arithmetic. The function that is perhaps more important than any other—the identity function assuming the value x for any x—lacks a traditional symbol. Therefore, in contrast to (4), in

$$\mathbf{D}\log x = 1/x, \; for \; x > 0, \quad \mathbf{D}x^3 = 3x^2, \quad \mathbf{D}\left(\tfrac{1}{2}x^2\right) = x, \qquad (6)$$

x cannot be shed. One obviously would not write

$$\mathbf{D}\log \; = 1/ \;, \; for \; x > 0, \quad \mathbf{D}^{\;3} = 3^{\;2}, \quad \mathbf{D}\left(\tfrac{1}{2}^{\;2}\right) = \;,$$

and this is why, for the sake of uniformity, x is also retained in (4). Traditionally, the identity function is referred to by its value for x as "the function x"—incidentally, a fourth current meaning of the italic letter x. There is, however, another way to preserve uniformity, namely, by rescuing the identity function from anonymity. If one denotes it by *j* (so that *j*x = x and *j*ⁿ x = xⁿ for any x), then (6) reads

$$\mathbf{D}\log x = j^{-1}x \; for \; x > 0, \; \mathbf{D}j^3x = 3j^2x, \; \mathbf{D}\left(j^2\right)x = x; \qquad (\text{KM:6*})$$

and in (6*), x can be shed as in (4):

$$\mathbf{D}\log = j^{-1} \; (\text{on the class P of all positive numbers}), \qquad (6')$$

$$\mathbf{D}j^3 = 3j^2, \; \mathbf{D}\left(\tfrac{1}{2}j^2\right) = j.$$

The formulas thus obtained (or, if one pleases, the results of calculus "in this notation") connect the functions themselves rather than their values and are of algebraic beauty—as it were, streamlined. Yet it is a matter of taste which one prefers: (4') and (6') or (4) and (6). But it is a fact (proved at the end of Chapter IV) and not a mere matter of taste that only if the identity function is granted the same status that log and cos have enjoyed for centuries can calculus be (to use another term that is in vogue) completely automatized. Automation is impossible "in the classical notation." In the automatized calculus one obtains specific results from general theorems by replacing function variables by the designation of specific functions, and numerical variables by the designations of specific numbers. For instance, from (5) one obtains:

$$\text{If } \mathbf{D}\,log = j^{-1}, \text{ then } \int_a^b j^{-1} = \Big[log\Big]_a^b$$

Since, according to (6'), actually $\mathbf{D}\,log = j^{-1}$ on the class P, one concludes that $\int_a^b j^{-1} = \Big[log\Big]_a^b$, where the numerical values a and b maybe replaced by any two numbers in the class P. For instance

$$\int_1^2 j^{-1} = \Big[log\Big]_1^2 \ (= log\,2 - log\,1 = log\,2).$$

A symbol for the identity function was introduced, as early as 1924, by the Russian logician Schönfinkel and has been adopted by H. B. Curry in his work on combinatoric logic. But whereas in topology extensive use has been made of a symbol for the identity mapping, that for the identity function had not, before the publication of the booklet "Algebra of Analysis" (see Bibliography), found its way into mathematical analysis—in particular not even into Curry's own mathematical papers.

To the specialist, perhaps more interesting is the automation of the theory of what, traditionally, is called the theory of functions of two variables. Most important are the functions assuming for every pair (x, y) the values x and y. Traditionally, they are referred to by their values as "the function x"—a fifth meaning of x—and "the function y." In an automatized calculus, they cannot remain anonymous. In Chapter XI they are called the selector *functions* and are denoted by I and J so that

$$I(x,y) = x \ \text{ and } \ J(x,y) = y \text{ for any x and y.}$$

Chapter XI contains the theories of partial derivatives and partial integrals. Calculus thus developed as an algebra of functions synthesizes the spirit of the oldest logic and that of the most modern machine age.

Derivative of a composite function

Let f and g be functions. The substitution of the second into the first, a process known as composition of two functions and usually denoted by the symbol $f \circ g$, is represented in Menger's book by juxtaposition of the two symbols. In this notation, $(fg)(x) = (fg(x))$ and $(\mathbf{D}f)g$ is the result of substituting g into $\mathbf{D}f$. The derivative of the function fg is therefore written as

$$\mathbf{D}(fg) = (\mathbf{D}f)g \cdot \mathbf{D}g.$$

The dot cannot be omitted in products of functions (in order to avoid confusion of a product with composition). Menger gives an exceptionally clear graphic argument for reaching the above result [30, pp. 233–7]. In an informative and well-written article [182], Erber has summarized Menger's approach and discussed its applications to physics.

Calculus without differentials

Chapter 8 of *CalcAMA* is entitled "Calculus of Derivatives", which leaves no room for doubt, for the book deals with derivatives and antiderivatives, and differentials are mentioned only as a relic of history and a testament to the inertia of the human mind. Needless to say, Menger's notation for antiderivatives made no use of differentials, but it seems necessary to add that he seems to have been unaware that the idea did not originate with him. As has been mentioned before, a few British authors—including West, Whewell and Airy—and at least three distinguished American mathematicians, namely W. E. Byerly [64, 65], William Ferrel [175], and E. B. Wilson [68] had already used Airy's symbol. This particular aspect of Menger's work was highlighted in an article that is discussed in the next section.

A zealous supporter

One of Menger's staunchest supporters, Philip S. Marcus (abbreviated hereafter as PSM), published a paper [183] with the intention of presenting the salient features of Menger's formulation of calculus, and pointing out the faults of the traditional approach (that is, of H-Calculus). The purpose of this section is to comment on PSM's paper. Although the article is not divided into sections, I will insert the symbol § and page number(s) to facilitate cross referencing to various parts of the text.

242

Introductory remarks [p. 324]

§1. The desirability of a variable-free notation for calculus may be illustrated by the following simple argument that $1 = 2$. Let $z = z(x,y)$, $x = x(r,s)$, $y = y(r,s)$. Then by the appropriate chain rule,

$$\frac{\partial z}{\partial r} = \frac{\partial z}{\partial x}\frac{\partial x}{\partial r} + \frac{\partial z}{\partial y}\frac{\partial y}{\partial r}$$

Cancel differentials (ignoring futile appeals that partial derivatives are not fractions, an edict which is plainly contradicted by notation and which in practice is disregarded by most physical chemists).

$$\frac{\partial z}{\partial r} = \frac{\partial z}{\partial x}\frac{\partial x}{\partial r} + \frac{\partial z}{\partial y}\frac{\partial y}{\partial r} = 2\frac{\partial z}{\partial r}$$

Since in general $\partial z/\partial r \neq 0$, we may safely conclude that $1 = 2$.

Those who have gone through the previous chapters will be able to see the flaws in the case prepared by PSM. Accordingly, I move on to a different issue.

Integration without differentials

PSM mentions three principles on which Menger constructed his variable-free calculus, and writes (p. 235):

[The abolition of differentials] has the most dramatic impact on traditional usage and meets with the most solid resistance. This resistance probably stems from three sources:

(a) the erroneous belief that the evaluation of integrals like $\int_0^1 (1 - x^2)^{1/2}\,dx$ or differential equations such as $dy/dx = xy$ require techniques based on differentials.

(b) the use of differentials as place-indicator, as in $\partial xy/\partial x$, $\partial xy/\partial y$, $\int xy\,dx$, $\int xy\,dy$, $\int xy\,ds$.

(c) popularity of differentials among physicists and other scientists.

To look at what PSM says about the role of substitution in the evaluation of integrals, we go to p. 326:

§2. Now let us consider the evaluation of integrals and differential equations. Let us start with a simple integral such as $\int x(1 - x^2)^{1/2}\,dx$.

It used to be obligatory to evaluate this integral by a so-called change of variable $u = 1 - x^2$. The magic formula $du = -2x\,dx$ should, but does not, lead to the incorrect equations

$$\int x(1-x^2)^{1/2}\,dx = -\frac{1}{2}\int \sqrt{u}\,du = -\frac{u^{3/2}}{3} + c$$

but a little hand-waving to the effect that u in the second integral is not really a so-called 'dummy' variable since it carries the important message $u = 1 - x^2$, gives us the correct answer (apart from correctness of notation)

$$= -\frac{1}{3}(1-x^2)^{3/2} + c,$$

where, as is traditional, functions are confused with variables and 'x' stands for the identity function.

§3. Nowadays, it has become respectable to recognize that $\int x(1-x^2)^{1/2}\,dx$ has the form $\int x\sqrt{g(x)}$ where $g(x) = 1 - x^2$, and to use the chain rule to find an anti-derivative by

$$\int x(1-x^2)^{1/2}\,dx = -\frac{1}{2}\int g'(x)(g(x))^{1/2}\,dx = -\frac{(g(x))^{3/2}}{3} + c.$$

[p. 327] Here the differentials appear, but play absolutely no role, since the necessary work is done by derivatives and the chain rule. In general, differentials can always be avoided if differentiation and especially the chain rule are thoroughly understood. For example, the above problem in Menger notation reads

$$\int j \cdot (1-j^2)^{1/2} = -\frac{1}{2}\int D(1-j^2) \cdot (1-j^2)^{1/2} = -\frac{(1-j^2)^{3/2}}{3} + c$$

and differentials have been completely abolished.

Excursus

The argument given in §2 (which concerns d-Calculus) simply does not hold water, as may be seen by repeating the traditional treatment without using the substitution $u = 1 - x^2$:

$$\int x(1-x^2)^{1/2}\,dx = -\frac{1}{2}\int (1-x^2)^{1/2}\,d(1-x^2)$$

$$= -\frac{(1-x^2)^{3/2}}{3} + c.$$

The above derivation is to be compared with that offered at the end of §3.

Since it is hard to decide whether the point made in §3 applies to d-Calculus or to H-calculus, I will examine both possibilities. Within the framework of d-Calculus, PSM has changed the variable of integration from x to $g(x)$, but differentials are an indispensable part of the argument here, since one uses the fact that $g'(x)\,dx = d(g(x))$, so that

$$\int x(1-x^2)^{1/2}dx = \int (1-x^2)^{1/2}\,x\,dx = -\tfrac{1}{2}\int (g(x))^{1/2}g'(x)\,dx$$

$$= -\tfrac{1}{2}\int (g(x))^{1/2}\,d(g(x)) = -\frac{(g(x))^{3/2}}{3} + c$$

If the argument is supposed to apply to H-Calculus, then the first displayed equation in §3 makes little sense, because each integral ends with dx (meaning that x is the variable of integration in both cases, and the second integral is just as hard (or easy) to find as the first. Let us rephrase the first equality in PSM's derivation

$$\int (1-x^2)^{1/2}\,x\,dx = \int [g(x)]^{1/2}\left[-\tfrac{1}{2}D_xg\right]dx$$

The trick is to use next Eqs. (2.19) and (2.22), which gives

$$-\tfrac{1}{2}\int [g(x)]^{1/2}\cdot D_xg\,dx = -\tfrac{1}{2}\int [g(x)]^{1/2}\,dg,$$

in which dx and dg are not differentials but signs indicating the variable of integration; needless to say, the g-integral on the right-hand side can now be deduced immediately.

Some symbol for indicating a change of variable is vital, but it need not be the highly confusing symbol used in H-Calculus. We will now make use of Airy's notation and the recipe given in Eqs. (2.23) and (2.26) to carry out the required integration. With $u = 1 - x^2$, one gets

$$\int_x x(1-x^2)^{1/2} = \int_u (1-u)^{1/2}u^{1/2}\cdot D_ux \qquad (15.1)$$

$$= \int_u (1-u)^{1/2}u^{1/2} \cdot \left(-\frac{1}{2(1-u)^{1/2}}\right) \quad (15.2)$$

$$= -\frac{1}{2}\int_u u^{1/2} = -\frac{1}{3}u^{3/2} + c \quad (15.3)$$

$$= -\frac{(1-x^2)^{3/2}}{3} + c. \quad (15.4)$$

For the sake of readers who are more comfortable with the notation of H-Calculus, the argument is repeated below (without changing the numerical labels of the equations), but—in order to block the cancellation of differentials—$D_u x$ is used instead of dx/du

$$\int x(1-x^2)^{1/2}\,dx = \int (1-u)^{1/2}u^{1/2} \cdot D_u x\,du \quad (15.1^*)$$

$$= \int (1-u)^{1/2}u^{1/2} \cdot \left(-\frac{1}{2(1-u)^{1/2}}\right) du \quad (15.2^*)$$

$$= -\frac{1}{2}\int u^{1/2}\,du = -\frac{1}{3}u^{3/2} + c \quad (15.3^*)$$

$$= -\frac{(1-x^2)^{3/2}}{3} + c. \quad (15.4)$$

We return now to the article written by PSM.

Two integrals [p. 327]

§4. Now let us consider a more difficult integral $\int_0^1 \sqrt{(1-x^2)}\,dx$. The so-called change of variables traditionally used here, $x = \cos u$, is not a simplification at all but a deliberate complication. To evaluate this integral in Menger notation, one must first have used the chain rule to prove the following theorem:

If g is one-to-one on the interval $[c,d]$ and $g(c) = a$, $g(d) = b$ then for any f continuous on the interval $[a,b]$

$$\int_a^b f = \int_c^d (f \circ g) \cdot Dg$$

(the method of substitution).

Then

$$\int_0^1 \sqrt{1-j^2} = \int_0^{\pi/2} \sqrt{1-\sin^2} \cdot D\sin \qquad (15.5a)$$

$$= \int_0^{\pi/2} \cos^2 \qquad (15.5b)$$

$$= \int_0^{\pi/2} \frac{1+\cos \circ 2j}{2} \qquad (15.5c)$$

$$= \tfrac{1}{2}j\left(\frac{\pi}{2}\right) + \tfrac{1}{4}\sin \circ 2j\left(\frac{\pi}{2}\right) - \tfrac{1}{2}j(0) - \tfrac{1}{4}\sin \circ 2j(0) \qquad (15.5d)$$

$$= \frac{\pi}{4}. \qquad (15.5e)$$

Excursus

To do this integration in D-Calculus, it will be enough to deal with the indefinite integral. We put $x = \sin u$, which implies that $\sqrt{1-x^2} = \cos u$ and $D_x u = \cos x$, and write

$$\int_x \sqrt{1-x^2} = \int_u \cos u \cdot D_x u \qquad (15.6)$$

$$= \int_u \cos^2 u = \int_u \frac{1+\cos 2u}{2} \qquad (15.7)$$

To convert the last three equations to their counterparts in H-Calculus, the reader should

$$\text{replace} \quad \int_\mu (\cdots) \quad \text{by} \quad \int (\cdots)\, d\mu, \qquad (\mu = x \text{ or } u). \quad (15.8)$$

It may not be amiss here to recall John West's approach to integration (see Chapter 14). West, or anyone following him, would integrate, not the function $\sqrt{1-x^2}$, but $\sqrt{1-x^2} \cdot x'$, where the prime denotes differentiation with respect to some unspecified variable, and integration is also to be performed with respect to the same variable. The other steps are shown below:

$$\int \sqrt{1-x^2} \cdot x' = \int \sqrt{1-x^2} \cdot \frac{x'}{u'} \cdot u' \qquad (15.9)$$

$$= \int \cos^2 u \cdot u' = \tfrac{1}{2} \int (1+\cos 2u) \cdot u' \quad (15.10)$$

At this point one would choose u as the variable of integration, which means that now $u' = 1$, but its presence helps specify the variable of integration.

We return now to the final part of PSM's article (pp. 327–8), where he discusses the solution of differential equations.

Solving a differential equation

§5. Now let us consider the differential equation $dy/dx = xy$. In Menger notation, there is, as usual, nothing more involved in solving this than the chain rule. [p. 328] The problem, in Menger notation, reads

$$Df = j \cdot f \qquad (15.11)$$

then by ordinary algebra,

$$\frac{Df}{f} = j \qquad (15.12)$$

so that by the chain rule,

$$\log f = \frac{j^2}{2} + c \qquad (15.13)$$

$$= K\exp\frac{j^2}{2} \qquad (15.14)$$

or, returning to a Cartesian representation,

$$y = K\exp(x^2/2). \qquad (15.15)$$

The above examples show that differentials are not needed for any computations in calculus. But what about the logical role played by differentials as place-indicators? It is a consequence of Menger's analysis of the concept 'variable' that functions are not variables. The historical confusion between functions and 'dependent variables' is regarded by Menger as a part of the historical evolution of the abstract function concept. Just as the abstract concept of number evolved only gradually from the concept of a number of chickens or a number of goats or a number of pebbles, so the abstract concept of function is still emerging from the concept of a function of time or a function of temperature or a function of x.

Excursus

In D-Calculus, one would begin by dividing the differential equation $D_x y = xy$ by y, and then use the chain rule to convert the left-hand side of the resulting equation into an expression that can be written as $D_x(\cdots)$

$$(1/y) \cdot D_x y = D_y(\ln y) \cdot D_x y = D_x(\ln y) \qquad (15.16)$$

Therefore $D_x(\ln y) = x$, which implies that $\ln y = x^2/2 + c$, and leads to Eq. (15.15).

Appendices

Appendix A

Differentiability, Differentials and Derivatives

A LTHOUGH the reader is supposed to be familiar with all the basic concepts of calculus, the topics of differentiability and differentiation of a composite function are usually not treated adequately in most elementary textbooks. A discussion of these topics is useful for gaining acquaintance with the term *principal part* and for deriving the chain rule for derivatives.

The differentiability of a function

The function $y = f(x)$ is said to be differentiable at the point x_0 if its increment Δy at this point can be represented as

$$\Delta y = A\,\Delta x + \Omega(\Delta x)\,\Delta x, \qquad (A.1)$$

where A is a number independent of Δx and $\Omega(\Delta x)$, a function of the argument Δx, has the property that

$$\Omega(\Delta x) \xrightarrow[\Delta x \to 0]{} 0, \quad i.e. \lim_{\Delta x \to 0} \Omega(\Delta x) = 0. \qquad (A.2)$$

Let us establish the relationship between the differentiability of a function at a point and the existence of derivative at that point by proving the following theorem.

251

Theorem 1. For a function $y = f(x)$ to be differentiable at the point x_0, it is necessary and sufficient that the function should possess a finite derivative at this point.

Proof. To prove the first part (necessity), we must show that if the function $f(x)$ is differentiable at the point x_0, $f'(x_0)$ exists. Now, if the function is differentiable at x_0, its increment at this point can be represented by Eq. (A.1). Dividing both sides of this equation by Δx (with $\Delta x \neq 0$), one gets

$$\frac{\Delta y}{\Delta x} = A + \Omega(\Delta x), \tag{A.3}$$

which gives, when one passes to the limit $\Delta \to 0$

$$\lim_{\Delta x \to 0} \frac{\Delta y}{\Delta x} = \lim_{\Delta x \to 0} \left(A + \Omega(\Delta x) \right) = A. \tag{A.4}$$

This proves that that the derivative $f'(x_0)$ exists at this point and equals A. Thus differentiability implies the existence of a derivative.

To prove the sufficiency, we must demonstrate that the existence of a finite derivative $f'(x_0)$ leads to Eq. (A.1) and (A.2). Let A denote $f'(x_0)$, and introduce the function

$$\Omega(\Delta x) = \frac{\Delta y}{\Delta x} - A. \tag{A.5}$$

Eq. (A.5) can be transformed to Eq. (A.1). Since the right-hand side of Eq. (A.5) tends to 0 as $\Delta x \to 0$, it follows that

$$\Omega(\Delta x) \xrightarrow[\Delta x \to 0]{} 0, \text{ in agreement with Eq. (A.2)}. \tag{A.6}$$

It will be noted that the function $\Omega(\Delta x)$ introduced in Eq. (A.5) is not defined for $\Delta x = 0$. By the same token, the right-hand side of Eq. (A.1) is also not defined for $\Delta x = 0$. Eq. (A.5) can be made applicable even at $\Delta x = 0$, by assigning an arbitrary value to $\Omega(0)$, and it is customary to take this value to be the limiting value, or $\Omega(0) = \lim_{\Delta x \to 0} \Omega(\Delta x) = 0$.

Principal part

It is possible to show [184, p. 131], without resorting to calculus, that $\sin x$ and $\cos x$ can each be expanded as an infinite series:[1]

$$\sin x = x - \frac{x^3}{3!} + \frac{x^5}{5!} + \cdots \tag{A.7}$$

$$\cos x = 1 - \frac{x^2}{2!} + \frac{x^4}{4!} - \cdots \tag{A.8}$$

Dividing both sides of Eq. (A.7) and taking the limit $x \to 0$ gives

$$\lim_{x\to 0} \frac{\sin x}{x} = \lim_{x\to 0} \left[1 - \frac{x^2}{3!} + \frac{x^4}{5!} + \cdots\right]$$

$$= 1 \tag{A.9}$$

One is often interested in the limiting behaviour (rather than the limiting value) of an expression. With this in mind, let us return to Eq. (A.7), and notice that, if one's interest lies in inferring only the limiting behaviour of $\sin x$, it is not absolutely necessary to divide Eq. (A.7) by x before one lets $x \to 0$. One is entitled to drop terms involving powers of x higher than the first on the right-hand side of Eq. (A.7), which immediately leads to the conclusion that

$$\sin x \xrightarrow[x\to 0]{} x. \tag{A.10}$$

Similarly we can write

$$\cos x \xrightarrow[x\to 0]{} 1, \quad 1 - \cos x \xrightarrow[x\to 0]{} \tfrac{1}{2}x^2. \tag{A.11}$$

In this context, it is convenient to use the "small o" notation. The equation $y = o(x)$ will be used to indicate that as $x \to 0$,

[1]The author (E. W. Hobson) pointed out in his Preface: "A strict proof of the expansions of the sine and cosine of an angle in powers of the circular measure has been given in Chapter VIII; this is a case in which, in many of the text books in use, the passage from a finite series to an infinite one is made without any adequate investigation of the value of the remainder after a finite number of terms, simplicity being thus attained at the expense of rigour."

$|y|/|x| \to 0$. Using this notation, we will write

$$\sin x = x + o(x), \quad 1 - \cos x = \tfrac{1}{2}x^2 + o(x^2),$$

and call the first term on the right-hand side of each equation as the *principal part* of the function on the left-hand side. Similarly, we may express the contents of Eq. (A.1) as

$$\Delta y = D_{xy}\Delta x + o(\Delta x)$$

say that Δy is the sum of its principal part, $D_{xy}\Delta x$ and an in-finitesimal of a higher order.

Derivative of the function of a function

If $y = f[g(x)] = \phi(x)$, ϕ is said to be a composite function, and denoted by the symbol $f \circ g$.

Let us consider the following pair:

$$y = f(u), \tag{A.12a}$$
$$u = g(x). \tag{A.12b}$$

In Eq. (A.12a) u is an independent variable, but it becomes a dependent variable in Eq. (A.12b). It is assumed that we are able to differentiate y with respect to u and also differentiate u with respect to x, and we wish to express $D_x y$ in terms of $D_u y$ and $D_x u (\neq 0)$.

To deduce the required relation, we follow Sargent [185] and begin with

$$\Delta u = (D_x u + \Omega_1)\,\Delta x, \text{ and} \tag{A.13}$$
$$\Delta y = (D_u y + \Omega_2)\,\Delta u, \tag{A.14}$$

where $\Omega_1 \to 0$ as $\Delta x \to 0$ and $\Omega_2 \to 0$ as $\Delta u \to 0$. Substituting for Δu from Eq. (A.13) in Eq. (A.14), we find that Δy is given by an expression of the form

$$\Delta y = (D_u y \cdot D_x u + \Omega)\,\Delta x, \tag{A.15}$$

where $\Omega \to 0$ as $\Delta x \to 0$. Hence y is a differentiable function of x and

$$D_x y = D_u y \cdot D_x u. \tag{A.16}$$

It will be instructive to restate the chain rule after replacing y by $f(u)$ in Eq. (A.16), which gives

$$D_x f(u) = D_u f(u) \cdot D_x u, \tag{A.17}$$

a result that is worth stating also in the prime notation (where a prime indicates differentiation with respect to the argument):

$$D_x f(u) = f'(u) \cdot u'(x), \tag{A.18}$$

or $\qquad \phi'(x) = f'(u) \cdot u'(x). \tag{A.19}$

If we wish to express the chain rule entirely in terms of x, we denote $y = f(u)$ by $y = \phi(x)$ and write

$$\phi'(x) = f'[g(x)] \cdot g'(x). \tag{A.20}$$

As an equation involving *functions* rather than numbers, the chain rule may be expressed as follows:

$$D(f \circ g) = (Df \circ g) \cdot Dg. \tag{A.21}$$

If we set $y = x$ in Eq. (A.16), the left-hand side becomes $D_x x = 1$, and we get

$$1 = D_u x \cdot D_x u, \tag{A.22}$$

which means
$$D_u x = [D_x u]^{-1}. \tag{A.23}$$

This result gives a simple method of differentiating inverse functions, for $y = f[g(x)] = x$ implies $u = g(x) = f^{-1}(x)$ and, conversely, $x = f(u)$. Therefore

$$D_x u = D_x[f^{-1}(x)] = \frac{1}{D_u x} = \frac{1}{D_u[f(u)]} = \frac{1}{D_u f[f^{-1}(x)]}.$$

The treatment of the chain rule in many textbooks leaves much to be desired. To those who appreciate the pedagogic potential of well-drawn diagrams, I recommend Menger's exposition [30, p. 233–8]; a later article by Holden [186] uses essentially the same approach. Menger, who uses the name *substitution rule* concludes his discussion by presenting the traditional statement of the rule in the traditional form, namely

$$\frac{dw}{du} = \frac{dw}{dv} \cdot \frac{dv}{du},$$

and ends with the punch lines: This is the formulation of the Substitution Rule in Leibniz' notation. And every freshman who sees this formula says "Of course."

Appendix B

Notation and Typesetting

Among the worst of barbarisms is that of introducing symbols which are quite new in mathematical, but perfectly understood in common, language. Writers have borrowed from the Germans the abbreviation $n!$ to signify $1 \cdot 2 \cdot 3 \cdots (n-1)n$, *which gives their pages the appearance of expressing surprise and admiration that 2, 3, 4, &c. should be found in mathematical results.* (My italics.)

The subject of mathematical printing has never been methodically treated, and many details are left to the compositor which should be attended to by the mathematician. Until some mathematician shall turn printer, or some printer mathematician, it is hardly to be hoped that this subject will be properly treated.

Augustus De Morgan in *The Penny Cyclopædia* [187, p. 444]

IF the title of this book had been *Calculus Without Symbols*, you would have immediately concluded that it was written by a charlatan, and you would not have been reading this appendix. Some readers might still come to the same conclusion, and I would be pleased if their disappointment leads them to write their own books on calculus.

Whether or not you believe in the aphorism 'no pain, no gain', you cannot deny, after thinking about modern mathematics, the truth of 'no symbols, no maths'. A notation is more than merely a set of symbols and some rules for combining these symbols to obtain mathematical results and to express them neatly. A good notational system becomes a source of ideas by inviting extensions and generalizations. For an elaboration of the above comments,

257

the reader may turn to Chapter V (entitled 'The Symbolism of Mathematics') in Whitehead's *An Introduction to Mathematics* [79], or to other sources [188, 189].

Two excerpts from Whitehead

It seems worthwhile to reproduce below two passages from the abovenamed chapter [79, pp. 61–3]. The first describes the practical advantages which result from using a good notation:—

> Mathematics is often considered a difficult and mysterious science, because of the numerous symbols which it employs. Of course, nothing is more incomprehensible than a symbolism which we do not understand. Also a symbolism, which we only partially understand and are unaccustomed to use, is difficult to follow. In exactly the same way the technical terms of any profession or trade are incomprehensible to those who have never been trained to use them. But this is not because they are difficult in themselves. On the contrary they have invariably been introduced to make things easy. So in mathematics, granted that we are giving any serious attention to mathematical ideas, the symbolism is invariably an immense simplification. It is not only of practical use, but is of great interest. For it represents an analysis of the ideas of the subject and an almost pictorial representation of their relations to each other. If anyone doubts the utility of symbols, let him write out in full, without any symbol whatever, the whole meaning of the following equations which represent some of the fundamental laws of algebra:—
>
> $$x + y = y + x \tag{1}$$
> $$(x + y) + z = x + (y + z) \tag{2}$$
> $$x \times y = y \times x \tag{3}$$
> $$(x \times y) \times z = x \times (y \times z) \tag{4}$$
> $$x \times (y + z) = (x \times y) + (x \times z) \tag{5}$$
>
> Here (1) and (2) are called the commutative and associative laws for addition, (3) and (4) are the commutative and associative laws for multiplication, and (5) is the distributive law relating addition and multiplication. For example, without symbols, (1) becomes: If a second number be added to any given number the result is the same as if the first given number had been added to the second number.

This example shows that, by the aid of symbolism, we can make transitions in reasoning almost mechanically by the eye, which otherwise would call into play the higher faculties of the brain.

Whitehead goes on to point a second use of symbolism, namely that of opening new avenues of thought:—

This second symbolic use is at first sight so absurdly simple that it is difficult to make a beginner realize its importance. Let us start with a simple example. In Chapter II. we mentioned the correlation between two variable numbers x and y represented by the equation $x + y = 1$. This can be represented in an indefinite number of ways; for example, $x = 1 - y$, $y = 1 - x$, $2x + 3y - 1 = x + 2y$, and so on. But the important way of stating it is

$$x + y - 1 = 0.$$

Similarly the important way of writing the equation $x = 1$ is $x - 1 = 0$, and of representing the equation $3x - 2 = 2x^2$ is $2x^2 - 3x + 2 = 0$. The point is that all the symbols which represent variables, e.g. x and y, and the symbols representing some definite number other than zero, such as 1 or 2 in the examples above, are written on the left-hand side, so that the whole left-hand side is equated to the number zero. The first man to do this is said to have been Thomas Harriot, born at Oxford in 1560 and died in 1621. But what is the importance of this simple symbolic procedure? It made possible the growth of the modern conception of *algebraic form*.

A few more words about Thomas Harriot appear to be in order here. Although remembered today by few, he was a giant who deserves to be placed alongside Galileo and Kepler. Many mathematicians in the seventeenth century devised new notational systems, but Harriot and later Leibniz stand above all others [190]. Harriot tried to develop a notation that would make verbal explanations redundant, and he may be justifiably called the father of modern algebra [191, 192].

What is to follow

In order to find out a little more about the considerations that must be taken into account by someone who wants to add a new symbol or introduce a substantial change in an established notational system, it will be instructive to listen to an author who took particular pains to discuss the technical issues that demand careful attention. In the passage reproduced below, De Morgan sheds light on some of the problems which are encountered in typesetting a mathematical text.

De Morgan distilled his views on the choice of symbols in the article on 'Symbols and Notation' in *The Penny Cyclopædia* [187], and in a long essay entitled *Calculus of Functions*, which appeared first in 1836 [193] and later in *Encyclopædia Metropolitana* [194]. The article in *The Penny Cyclopædia* is reproduced in the following section in its entirety, with only one change in notation: De Morgan's symbol ϕx has been replaced by its modern equivalent, namely $\phi(x)$. References (within square brackets) to other articles in the same *Cyclopædia* have been left intact.

Symbols and notation

The word symbol (from the Greek 'symbolon,' $\sigma \acute{v} \mu \beta o \lambda o v$) means 'that which is taken with,' and a symbol is a mark which is always attached to some one particular meaning. Notation (*nota*, a known mark) is the method of selecting and assigning meaning to symbols, and the theory of notation (if it yet deserve the name) includes the consideration and choice of symbols,with the formation of rules of selection, so as to take the symbols which are best adapted for the purpose.

This subject might be treated in a very wide manner, for all marks with understood meanings are symbols, from written words to direction-posts. A picture is a symbol, the force of which lies in the resemblance to its object, and many of the earliest symbols must have been pictorial. It is obvious that a general treatment of the subject would hardly be within the power of any one person, and that its extent would be enormous, though it would be desirable to have it discussed in a more general form than has yet been attained, in order that its different parts may receive aid from the rest. Symbols are to the progress of civilization precisely what mechanism is to that of the arts, not a moving force, perfectly dead in themselves, but capable of being made the medium by which the power is conveyed to its destination, and adapted to its object. They are the instruments of our first thoughts and the originators of new ones. The process by which the earliest symbols called out a yet higher intelligence than that which produced them, which last was again employed in perfecting the symbols themselves, and so on alternately, exactly resembles what has taken place in the mechanical arts. The earliest and rudest tools were first employed to make better ones; and every improvement in the use of force has found one of its best applications in the construction of machinery itself.

We propose in this article to treat particularly of mathematical notation, which, like language, has grown up without much looking to, at the dictates of convenience and with the sanction of the majority. Resemblance, real or fancied, has been the first guide, and analogy has succeeded.

Signs are of two kinds,—1st, Those which spring up and are found in existence, but cannot be traced to their origin; 2ndly, Those of which we know either the origin, or the epoch of introduction, or both. Those of the first kind pass into the second as inquiry advances. [ALPHABET.] In our subject we have mostly to deal with the second class.

Mathematical marks or signs differ from those of written language in being almost entirely of the purely abbreviative character, since it is possible that any formula might be expressed in words at length. We say possible, because it is barely so, not meaning thereby to imply that the mathematical sciences could ever have flourished under a system of expressions in words. A well-understood collection of notions, however extensive, becomes simple as a matter of conception by use and habit, and thus becomes a convenient resting-point for the mind and a suitable basis for new combinations of ideas. Now it is the characteristic of the advance of human knowledge that the mind never grapples at once with all that is contained in the notions under use for the time being, but only with some abstraction derived from a previous result, or some particular quality of that result. Hence no symbol which should contain the representative of every idea which occurred in the previous operations would ever be necessary; and more than this, it would even be pernicious from its complexity, as also from its suggesting details which are not required. That generalization, or rather abstraction, which is the distinctive character of the civilized language as compared with the savage (though the latter is not wholly without it), must be the ruling process of mathematical notation, as it is of the ad-

262

vance of spoken language; and in this point of view the connection of our subject with speech presents more analogies and gives more instruction than its comparison with the written signs of speech. The latter is a bounded subject. When once it is agreed how the different modifications of sound shall be represented, written language follows immediately; nor do the infinite modes of using words require any modification of the method of writing them. In our modern works for instance, it would be difficult to find many artifices of notation with which to compare the never-ceasing varieties of mathematical signs. In mentioning the marks of punctuation and reference, the italics for emphatic words, and the varieties of print by which notes are distinguished from text, &c., we have almost exhausted the list.

The greatest purposes of notation seem to be answered when the reader or learner can tell what is meant, first, with the utmost certainty, secondly, with sufficient facility; it being always understood that the second must be abandoned when it clashes with the first. Too much abbreviation may create confusion and doubt as to the meaning; too little may give the meaning with certainty, but not with more certainty than might have been more easily attained. Thus the old algebraists, in using A *quadratum* for A multiplied by A, in their transition from words at length to simple notation, used ten symbols where two only are requisite; and those who first adopted the symbol A A lost no certainty, and gained materially in simplicity. The successors of these again, who employed AA, AAA, AAAA, &c., to stand for the successive powers of A, were surpassed in the same manner by those who adopted A^2, A^3, A^4, &c. Beyond this it is obvious the notation cannot go in simplicity. The symbol which is to represent 'n As multiplied together' must suggest all three components of the preceding phrase—namely, n and A, and multiplied together. In A^n, the n and A are obvious, and the position of the letters is the symbol of multiplication; but, on the other hand, those who teach the beginner to signify by A^2 the square described on the line A purchase simplicity at the expense of certainty. The same mathematical phrase with them stands for two different things, connected indeed, but of more dangerous consequence from that very connection; for where similarities exist the reader should not be made to convert them into identities. It is of as much importance to impress the distinction of the things signified, as the analogy of their properties.

Certainty, then, and the greatest facility of obtaining it, seem to be the main points of good notation; and this is true with respect to the learner of all that has gone before. Grant that the mathematical sciences are never to advance further, and many alterations might be made, and many new practices adopted, which would give facility in acquiring the past, without any introduction of obscurity. But the future must also be thought of; and no scheme will merit approbation which enlightens one end of the avenue at the expense of the other. Notation influences discovery by the suggestions which it makes: hence it is desirable that its suggestions should be as many, as plain, and as true, as it is possible. Here we are on quite a different ground: reason is the builder and settler, but imagination is the discoverer; and it might turn out that a notation which suggests many and obvious new ideas, though some of them should be fallacious, would be preferable in its consequences to another of less suggesting power, but more honest in its indications. And while we speak of positive suggestion, it must not be forgotten that a notation may be faulty in occupying the part of the symbol which properly belongs to the extension of another notation. The latter is thus deprived of its natural direction of growth; and must find its way elsewhere, to the injury perhaps of some other part of the symbol. In throwing together a few rules, previously to a little description of the present state of mathematical notation, we do not pretend to have exhausted the list of cautions which the subject requires. It is to be remembered that the language of the exact sciences, instead of being, as should be the case, a separate subject, is hardly ever treated at all, and then only in connexion with some isolated parts of the system. With the exception of an article by Mr. Babbage, in the Edinburgh Encyclopædia, we do not know of anything written in modern times on notation in general. Much may be collected, having notation for its specific object, from the writings of Arbogast, Babbage, Carnot, Cauchy, J. Herschel, and Peacock; writers who all have considered it necessary, when proposing a new symbol or modification of a symbol, to assign some reason for the proposal. In general, however, it is the practice to adopt or reject notation without giving any justification of the course pursued. If it could be rendered necessary, by the force of opinion, that every author should, in making a new symbol, explain the grounds, firstly, of his departure from established usage, secondly, of his choice from among the different methods which would most obviously present themselves, two distinct advantages would result. In the first place, we should in most cases retain that which exists, until something was to be gained by altering it: in the second, research and ingenuity would have a call into action which does not now exist. We hardly need mention a thing so well known to the mathematician as that the progress of his science now depends more than at any previous time upon the protection of established notation, when good, and the introduction of nothing which is of an opposite character. We should rather say the rate of progress; for, however bad may be the immediate consequences of narrow and ignorant views in this respect, they cannot be permanent. The language of the exact sciences is in a continual state of whole some fermentation, which throws up and rejects all that is incongruous, obstructive, and even useless. Had it been otherwise, it is impossible that the joint labours of three centuries and many countries, of men differing in language, views, studies, and habits, could have produced so compact and consistent a whole, as, with some defects (though no two persons agree precisely what they are), the present structure of mathematical language must be admitted to present.

The following rules and cautions, with respect to notation, are drawn from observation of the present state just alluded to.

1. Distinctions must be such only as are necessary, and they must be sufficient. For instance, in so simple a matter as the use of capitals or small letters, whatever may guide the inquirer to adopt either in one case should lead him to the same in another, unless some useful distinction can be made by the change. Thus a writer who in one instance uses a capital letter to denote a complicated function of small letters (which is a very desirable mode), will in another part of the same question employ a small letter for a similar purpose, thus nullifying an association of ideas which perspicuity would desire to be retained. If such a course were necessary in the first case, it is still more so in the second. It is not often that the second part of this rule is infringed; so small

an addition makes a sufficient distinction, that the principal danger which arises is that of the same notative difference occurring in two varied senses in *different* problems.

The tendency to error is rather towards over-distinction than the contrary. It is surprising how little practice enables the beginner in mathematics to remember that so slight a difference as that of a and a' implies two totally different numbers, neither having any necessary connexion with the other. The older mathematicians [ACCENT] overdid the use of distinctions in their uniform adoption of different and unconnected letters, and forgot resemblances.

2. The simplicity of notative distinctions must bear some proportion to that of the real differences they are meant to represent. Distinctions of the first and easiest order of simplicity are comparatively few; the complications of ideas of which they are the elements of representation are many, and varied to infinity. There is no better proof of skill than the adaptation of simple forms to simple notions, with a graduated and ascending application of the more complicated of the former to the more complicated of the latter. But some writers remind us in their mathematical language of that awkward mixture of long and short words to which the idiom of our language frequently compels them in their written explanations of the formulae. For example, if there be two words of more frequent occurrence than any others, they are *numerator* and *denominator*; the parts of a fraction cannot be described under nine syllables. A mathematician will have occasion to write and speak these words ten thousand times, for every occasion on which he will have to use the word cusp, of four letters. A comparatively rare idea, used in an isolated subject, can be expressed in one syllable, while the never-ending notions of the parts of a fraction require nine: this he cannot help; but it is in his power to avoid the same sort of inversion in his notation.

3. Pictorial or descriptive notation is preferable to any other, when it can be obtained by simple symbols. Many instances occur in astronomy, and the use of the initial letters of words may be cited as a class of examples: as in f for force, v for velocity, &c.

4. Legitimate associations which have become permanent must not be destroyed, even to gain an advantage. The reason is, that the loss of facility in reading established works generally

266

more than compensates for the advantage of the proposed notation; besides which, it seldom happens that the desired object absolutely requires an invasion of established forms. For instance,
perhaps the most uniform of all the notations of the higher mathematics is the use of the letter d to signify an increment which is
either infinitely small, or may be made as nearly so as we please.
A few Cambridge writers have of late years chosen to make a
purely arbitrary change, and to signify by dy, dx, &c., not increments, but limiting ratios of increments: and students trained in
these works must learn a new language before they can read Euler, Lagrange, Laplace, and a host of others. Thus $d_x y$ has been
made to stand for $dy : dx$, and the old association connected with
dy has (in the works spoken of) been destroyed. Now if the letter
D had been employed instead, the only harm would have been
that the student would have had to learn a new language before he communicated with the greatest mathematicians; as it is,
many will have to form a new language out of the materials of
the old one, which is a much harder task.

5. Analogies should not be destroyed, unless false: for true
analogy has been frequently the parent of discovery, and always
of clearness. Thus the analogy of $\sum \phi(x)\Delta x$ and $\int \phi(x)\,dx$ was
lost to the eye by the use of $\int_x \phi(x)$ to signify the latter; an innovation which preceded the one last mentioned, and has obtained
more approbation in this country. The notation used by Fourier
to express a definite integral, $\int_a^b \phi(x)\,dx$, will certainly prevent
the spread of the one just alluded to; though this last itself is
chargeable with breach of analogy: for $\int^2 \phi(x)\,dx^2, \int^3 \phi(x)\,dx^3$,
&c., ought to represent the successive integrals of $\int \phi(x)\,dx$. Fortunately, however, the symbols $(\int dx)^2\phi(x), (\int dx)^3\phi(x)$, &c., may
represent these successive integrals; and thus the two notations
may be combined. For instance, $(\int_0^x dx)^4\phi(x)$ represents the fourth
integral of $\phi(x)$, each integration being made from 0 to x.

6. False analogies should never be introduced; and, above
all, the incorrect analogies which custom and idiom produce in
language should not he perpetuated in notation. It is becoming rather common to make editions of Euclid which are called
symbolical, and which supply signs in the place of many words.
To this, if properly done, there cannot be any objection in point
of correctness: nor can we take any serious exception to the use

of □AB to stand for the square on AB, to $\|^{\ell}$ for parallel, $\angle$ for angle, $\perp^{r}$ for perpendicular, &c. But when we come to AB · BC for the rectangle on AB and BC, AB^2 for the square on AB, we feel the case to be entirely altered. These are already arithmetical symbols: it is bad enough that the word square should have both an arithmetical and a geometrical meaning, and causes plenty of confusion: a good notation, if it cannot help in avoiding this confusion, should at least not make it worse. At the same time, with regard to symbolic geometry, we feel some repugnance to introduce it into the elements, from observing that all the best writers seem to feel with one accord that pure reasoning is best expressed in words at length. If it be desirable that a student should be trained to drop reasoning, except as connecting process with process, and to think of process alone in the intervening time, it is also most requisite that he should have a corrective of certain bad habits which the greatest caution will hardly hinder from springing up while he is thus engaged. Arithmetic and algebra amply answer the first end; and geometry, in the manner of Euclid, is the correcting process. Will symbolic geometry do as well? We will not answer positively, but we must say we doubt it.

7. Notation may be modified for mere work in a manner which cannot be admitted in the expression of results which are to be reflected upon. The mathematical inquirer must learn to substitute, for his own private and momentary use, abbreviations which could not be tolerated in the final expression of results. Work may sometimes be made much shorter, and the tendency to error materially diminished, by attention to this suggestion.

For example, the complexity of the symbols,

$$\frac{dz}{dx}, \frac{dz}{dy}, \frac{d^2z}{dx^2}, \frac{d^2z}{dx\,dy}, \frac{d^2z}{dy^2},$$

greatly impedes the operations connected with problems in solid geometry: the letters p, q, r, s, t, which are often substituted for them, make us lose sight of the connection which exists between the meanings. But the symbols

$$z_x, z_y, z_{xx}, z_{xy}, z_{yy},$$

are not long nor complicated enough to partake much of the disadvantage of the complete symbols, while they are entirely free from that of the isolated letters.

8. In preparing mathematical writings for the press, some attention should be paid to the saving of room. In formulæ which stand out from the text, this is not of so much consequence; but in the text itself a great deal of space is often unnecessarily lost. For example, it is indispensable in formulæ to write a fraction, such as $\frac{a}{b}$ in the manner in which it here appears: but if this be done in the text, a line is lost; and, generally speaking, $a : b$, or $a \div b$, would do as well in mere explanation. Also, in printing, redundancies which are tolerated in writing, should be avoided, such as $\sqrt{7}$, where $\sqrt{7}$ would do as well.

9. Strange and unusual symbols should be avoided, unless there be necessity for a very unusual number of symbols The use of *script* letters, such as $\mathscr{A}$, $\mathscr{B}$ or old English letters, as 𝔄, 𝔅, 𝔞, 𝔟, &c., except in very peculiar circumstances, is barbarous. A little attention to the development of the resources of established notation will prevent the necessity of having recourse to such alphabets. Nor is it wise to adopt those distinctions in print which are not easily copied in writing, or which it is then difficult to preserve: such as the use of A and A, &c. in different senses; even the distinction of Roman and Italic small letters, a and a, &c., should be sparingly introduced.

10. Among the worst of barbarisms is that of introducing symbols which are quite new in mathematical, but perfectly understood in common, language. Writers have borrowed from the Germans the abbreviation $n!$ to signify $1 \cdot 2 \cdot 3 \cdots (n-1)n$, which gives their pages the appearance of expressing surprise and admiration that 2, 3, 4, &c. should be found in mathematical results.

The subject of mathematical printing has never been methodically treated, and many details are left to the compositor which should be attended to by the mathematician. Until some mathematician shall turn printer, or some printer mathematician, it is hardly to be hoped that this subject will be properly treated.

The elements of mathematical notation are as follows:—

1. The capitals of the Roman alphabet, and the small letters of the Italic. The small Roman letters and the Italic capitals are

rarely used, and should be kept in reserve for rare occasions.

2. The small letters of the Greek alphabet and such capitals as are distinguishable from the corresponding Roman ones, as Δ, Φ, Ψ.

3. The Arabic numerals, and occasionally the Roman ones.

Of all these there should be three different sizes in a good mathematical press, and the different sorts should bear a much better proportion to one another than is usual. The Greek letters seldom set properly with the Roman ones, and few indeed are the instances in which such symbols as

$$a^{m^n} \qquad e^{(1+i)^2}$$

are, as they ought to be, good copies of the manner in which they are written. The handwriting of a bad writer is frequently more intelligible to the mathematical eye than the product of the press. Among the faults to which the compositor is naturally subject, and which frequently remain uncorrected by the author, is that of placing blanks or spaces in the manner in which he would do in ordinary matter, by which he is allowed to separate symbols which are in such close connection that absolute junction would not be undesirable. For instance, $\cos \theta$ for $\cos \theta$, $(a\ b + c\ d)$ for $(ab + cd)$. As a general rule, the manuscript should be imitated.

4. Accents, superfixed and suffixed, as in $a''a_{\prime\prime}$. These are generally continued, when they become too numerous, by Roman numerals, as in $a, a_{\prime}, a_{\prime\prime}, a_{\prime\prime\prime}, a_{iv}, a_{v}, a_{vi}$, &c.

5. The signs $+$ $\times$ $\div$ $:$ $\sqrt{}$, and the line which separates the numerator from the denominator. Of these there are generally not sizes enough, particularly as to the sign . It frequently happens that such an expression as $(x - 1)(x - 2)(x - 3)$ &c. overruns a line very inconveniently, when the use of a shorter negative sign, as in $(x\text{-}1)(x\text{-}2)(x\text{-}3)$ would avoid such a circumstance altogether. Between the division line of a fraction and the numerator and denominator unsightly spaces very often occur, as in

$$\frac{a + b}{c + d} \quad \text{instead of} \quad \frac{a+b}{c+d}$$

6. The integral sign $\int$ with its limits expressed, as in $\int_a^b$ the symbols of nothing and infinity, 0 and ∞.

7. Brackets, parentheses, &c. [ ], ( ), { }, &c. These are often not properly accommodated to the size of the intervening expressions, particularly in thickness.

8. The signs of equality, &c., $=$, $<$, $>$.

9. Occasionally, but rarely, a bar or a dot is used over a letter, as $\bar{a}$ or $\dot{a}$. In some works, accents and letters are placed on the left of a symbol, as in $'a$, ^2a, $_2a$. This however should be avoided, as it is difficult to tell to which letter the symbol belongs; and there are ample means of expression in what has been already described.

There are no general rules laid down for the use of notation: a few hints however may be collected from the practice of the best writers of recent times.

1. When a letter is to be often used, it should be, if possible, a small letter, not a capital. The latter species is generally used for functions of small letters.

2. The letters d, Δ, δ, and D, are appropriated for operations of the differential calculus, and should hardly ever be used in any other sense.

3. When co-ordinates are used, the letters x, y, z, must be reserved to signify them; x, y, z, and ξ, η, ζ, may be used if different species be required, and if x', y', z', &c. or $x_{\prime}, y_{\prime}, z_{\prime}$, &c. should not be judged convenient.

4. When functional symbols are wanted, the letters ϕ, ψ, χ, F, f, Φ, Ψ, should be first reserved for them; afterwards ϖ, ω, η, sometimes π, ξ, μ, ν.

5. The letter π is, by universal consent, appropriated to 3. 14159..., and ε (by the French e) to 2. 71828...; Γ to the functional symbol for $1 \cdot 2 \cdot 3, \cdots n$.

6. When many operations of differentiation occur, superfixed accents should be avoided in any other sense than that of differentiation.

7. When exponents are wanted to aid in signifying operations, the powers should be carefully distinguished Thus, in a process in which $\sin^{-1} x$ is used for the angle whose sine is x, the square, cube, &c. of $\sin x$ should not be $\sin^2 x$, $\sin^3 x$, &c., but $(\sin x)^2$, $(\sin x)^3$, &c. Some writers, would have the latter notation employed in all cases; but this is, we think, asking a little too much.

8. Greek letters are generally used for angles, and Italic letters for lines, in geometry. To this rule it is desirable to adhere as far

as possible, but it cannot be made universal.

9. Suffixed numerals are generally the particular values of some function. Thus a_v, means a function of v, of which the values for $v = 0$, $v = 1$, &c. are a_0, a_1, &c.

10. As to the radical sign, $\sqrt{x}, \sqrt[3]{x}$, &c. do not generally mean any one of the square roots, cube roots, &c. of x, but the simple arithmetical root. The indeterminate root is usually denoted by the exponent. Thus $a \pm \sqrt{b}$ may be necessary, but $a \pm b^{\frac{1}{2}}$ has a superfluity.

11. The same letters should be used, as far as possible, in the same sense throughout any one work; and some preceding good writer should be followed. As a general rule, those only are entitled to invent new symbols who cannot express the results of their own investigations without them.

The writer who is most universally acknowledged to be a good guide in the matter of notation is Lagrange. This subject is of great importance; but fortunately it is pretty certain that no really bad symbol, or system of symbols, can permanently prevail. Mathematical language, as already observed, is, and always has been, in a state of gentle fermentation, which throws up and rejects all that cannot assimilate with the rest. A received system may check, but cannot ultimately hinder, discovery: the latter, when it comes, points out from what symbolic error it was so long in arriving, and suggests the proper remedy.

For the progress of mathematical language, see TRANSCENDENTAL; see also SYMMETRY.

Excerpt from De Morgan's *Encyclopædia Metropolitana* article

The occurrence of fractions, such as $\dfrac{a}{b}$, $\dfrac{a+b}{c+d}$ in the verbal part of mathematical works is a source of considerable loss of room, and creates an inelegant and even confused appearance in the printed page. It is very desirable, in every point of view, except the strictly mathematical one, that some method of representation should be adopted which does not require a larger space than is usual between two successive lines. At the same time, it is by no means of very great importance that the verbal part should entirely coincide with the mathematical part in notation,

272

so long as the latter remains to preserve the usual conventions. The symbol ÷ has been disused for a sufficient reason, namely,

reasons for not using ÷ the number of times which the pen must be taken off to form it. This has been, and we imagine always will be, the cause either of abandonment or abbreviation. The question is, whether a new and easy notation could not be substituted; and it is desirable that it should be derived from analogy, such as (accidentally, we believe) does exist in >, =, and <. If we look at × and +, and observe that the first is made by turning the second through half a right angle, denoting multiplication, which is primarily an extension of addition in like manner as division is an extension of subtraction, we may thus invent the symbol / or \ to denote di-

/: solidus vision, which is also the symbol of subtraction turned through half a right angle. If a/b were used to denote a divided by b, and $(a+b)/(c+d)$ to denote $a+b$ divided by $c+d$, all necessity for increased spacing would be avoided; but this alteration should not be introduced into completely mathematical expressions, though it would be convenient in particular cases.

G. G. Stokes

Stokes was among those who cared about typesetting niceties, and decided, when his mathematical and physical papers were to be published in a collection, to avoid the awkward uneven line spacing necessitated by the use of the symbol $\frac{dy}{dx}$, and he felt that the introduction of a new symbol could no longer be postponed [195, pp. vii–ix]:

> After full consideration, I determined to introduce an innovation in notation which was proposed a great many years ago, for at least partial use, by the late Professor De Morgan, in his article on the Calculus of Functions in the *Encyclopaedia Metropolitana* though the proposal seems never to have been taken up. Mathematicians have been too little in the habit of considering the mechanical difficulty of setting up in type the expressions which they so freely write with the pen; and where the setting up can be facilitated with only a trifling departure from existing usage as regards the appearance of the expression, it seems advisable to make the change.
>
> Now it seems to me preposterous that a compositor should be called on to go through the troublesome process of what printers

call *justification*, merely because an author has occasion to name some simple fraction or differential coefficient in the text, in which term I do not include the formal equations which are usually printed in the middle of the page. The difficulty may be avoided by using, in lieu of the bar between the numerator and denominator, some symbol which may be printed on a line with the type. The symbol ":" is frequently used in expressing ratios; but for employment in the text it has the fatal objection that it is appropriated to mean a colon. The symbol "÷" is certainly distinctive, but it is inconveniently long, and $dy \div dx$ for a differential coefficient would hardly be tolerated. Now simple fractions are frequently written with a slant line instead of the horizontal bar separating the numerator from the denominator, merely for the sake of rapidity of writing. If we simply consent to allow the same to appear in print, the difficulty will be got over, and a differential coefficient which we have occasion to name in the text may be printed as dy/dx. The type for the slant line already exists, being called a *solidus*.

On mentioning to some of my friends my intention to use the "solidus" notation, it met with a good deal of approval, and some of them expressed their readiness to join me in the use of it, amongst whom I may name Sir William Thomson and the late Professor Clerk Maxwell.

In the formal equations I have mostly preserved the ordinary notation. There is however one exception. It frequently happens that we have to deal with fractions of which the numerator and denominator involve exponentials the indices of which are fractions themselves. Such expressions are extremely troublesome to set up in type in the ordinary notation. But by merely using the solidus for the fractions which form the indices, the setting up of the expression is made comparatively easy, while yet there is not much departure from the appearance of the expressions according to the ordinary notation. Such exponential expressions are commonly associated with circular functions; and though it would not otherwise have been necessary, it seemed desirable to employ the solidus notation for the fraction under the symbol "sin" or "cos," in order to preserve the similarity of appearance between the exponential and circular functions.

In the use of the solidus it seems convenient to enact that it shall as far as possible take the place of the horizontal bar for which it stands, and accordingly that what stands immediately on the two sides of it shall be regarded as welded into one. Thus $\sin mrx/a$

means $\sin(mrx \div a)$, and not $(\sin mrx) \div a$. This welding action may be arrested when necessary by a stop: thus $\sin n\theta./r^n$ means $(\sin n\theta) \div r^n$ and not $\sin(n\theta \div r^n)$.

The only objection that I have heard suggested against the solidus notation on the ground of its being already appropriated to something else, relates to a condensed notation sometimes employed for factorials, according to which $x(x+a)\ldots$ to n factors is expressed by $x^{n|a}$ or by $x^{n/a}$. I do not think the objection is a serious one. There is no risk of the solidus notation, as I have employed it, being mistaken for the expression of factorials; of the two factorial notations just given, that with the separating line vertical seems to be the more common, and might be adhered to when factorials are intended; and if a greater distinction were desired, a factorial might be printed in the condensed notation as $x^{n(a}$, where the "(" would serve to recall the parentheses in the expression written at length.

Society for the Prevention of Cruelty to Printers

Stokes must have been delighted to get a nomination for the office of the president of a fictitious society [196, p. 397]:

> Dear Stokes,
>
> Best thanks for the Mathematical and Physical Papers. I think the *solidus* looks very well indeed and is really a great improvement: it would give you a strong claim to be President of a Society for the Prevention of Cruelty to Printers.[1] I ran through the volume to see whether you had adopted exp. x instead of e^x: you do not seem to have any exponents complicated enough to make this necessary or even advantageous. I think you do not approve of cosh and sinh.
>
> Believe me, yours very sincerely,
>
> A. Cayley

Many years were to pass before I discovered the inspiration for the name of the society which has been made redundant by the invention of LaTeX. If you like participating in harmless gossip about people who are all dead now, you might like to read Chapter IV (entitled "Society for the Prevention of Cruelty to Undergraduates") in Rouse Ball's *Cambridge Papers* [197].

---

[1] I have changed 'prevention' to 'Prevention'.

Appendix C

The Long and Short of the Integral Sign

Every schoolboy knows who imprisoned Montezuma, and who strangled Atahualpa. But we doubt whether one in ten, even among English gentlemen of highly cultivated minds, can tell who won the battle of Buxar, who perpetrated the massacre of the massacre of Patna, whether Sujah Dowlah ruled in Oude or in Travancore, or whether Holkar was a Hindoo or a Mussulman.

T. B. Macaulay [198, p. 1]

Anyone who has taken a first course in calculus know what the symbol $\int f(x)\,dx$ stands for. But we doubt whether one in ten, even among their teachers in the highly reputed colleges, can tell what the symbols $\int^x f(x)\,dx$ means, or who wrote the first book in the English language on Continental calculus, and what symbol he used instead of $\int f(x)\,dx$ and why its author spoke of *derivative equations* instead of *differential equations*.

The purpose of this appendix is to supply additional information concerning integration of functions and equations, and to comment on some symbols that have not yet been introduced.

The general antiderivative

Consider a determinate function $F(x)$, where by the term *determinate* we mean that the function does not contain a constant whose value is not known, and let its x-derivative be $f(x)$. Then the symbol

$$\int f(x)\, dx$$

is used to indicate the general antiderivative of the function $f(x)$ or the antidifferential of the differential $f(x)\, dx$ according as one is using H-Calculus or d-Calculus. Thus

$$\int f(x)\, dx = F(x) + C,$$

where C is a constant whose value remains indeterminate until some additional information is supplied.

This is all well and good, but speaking of the general antiderivative becomes exceedingly tedious, because it compels one to have clear ideas about *some* and *any*. Whitehead has written an entire chapter on this distinction [79, Ch. 2], and it is good to learn these things either before embarking on a study of calculus, or to think about them much later, or never. To convince the reader that, if we want to talk about general antiderivatives, there is no escape from the *some-any* distinction, I reproduce a passage from an excellent calculus book [101, pp. 209–10]:

> To find an antiderivative of the sum of two functions, find an antiderivative of each function and add these two antiderivatives. This rule is sometimes expressed in the form
>
> $$\int [f_1(x) + f_2(x)]\, dx = \int f_1(x)\, dx + \int f_2(x)\, dx, \qquad (1)$$
>
> but the interpretation of this formula calls for some special comment, in view of the earlier statement that $\int f(x)\, dx$ denotes the general antiderivative of f. A "general antiderivative" is not a single function, but a family of functions obtained by adding all possible constants to some particular antiderivative. How, therefore, are we to interpret (1), which in some way is supposed to express, not an equality of two functions, but an equality of two families of functions? There are various ways of giving a formal interpretation of (1); some ways involve more mathematical sophistication

than others. We shall put the matter as follows. Formula (1) is a short way of summing up two statements:

(a) *Every* antiderivative of $f_1 + f_2$ can be expressed by adding *some* antiderivative of f_1 and *some* antiderivative of f_2;

(b) the sum of *any* antiderivative of f_1 and *any* antiderivative of f_2 is *some* antiderivative of $f_1 + f_2$.

Other formulas which occur later and involve the symbol $\int$ more than once, or which involve the symbol $\int$ and the C denoting an arbitrary constant, are to be interpreted in a similar manner as statements asserting the equality of two families of functions.

Those who are interested merely in *using* calculus will never need to ponder over the general antiderivative. For such people, it is best to follow Augustus De Morgan and omit the constant of integration unless one is applying the result of an integration to a specific problem. We will first look at a long passage from his textbook [89, pp. 105–6], and discuss its implications afterwards:

Our first methods of integration must be the observation of differential coefficients, and the reconversion of each into an indefinite integral. Understanding always by $\int \phi(x)\,dx$ the integral with an arbitrary, but given, lower limit, and x itself for the higher limit, we see that if $\phi_1(x)$ differentiated gives $\phi(x)$, then $\int_a^x \phi(x)\,dx$ is $\phi_1(x) + C$. It is usual to omit the constant, as an attendant of the integral sign so well known that it is unnecessary except where we are actually applying the integral calculus, and may be dispensed with when we are merely ascertaining integral forms. We can thus find the following theorems:

1. $\int (u + v - w)\,dx = \int u\,dx + \int v\,dx - \int w\,dx.$

To prove that these are the same, observe that differentiated they give the same result. For $\dfrac{d}{dx}\int u\,dx = u$, consequently,

$$\frac{d}{dx}\int (u + v + w)\,dx = (u + v + w)$$

$$\frac{d}{dx}\left[\int u\,dx + \int v\,dx + \int w\,dx\right]$$
$$= \frac{d}{dx}\int u\,dx + \frac{d}{dx}\int v\,dx + \frac{d}{dx}\int w\,dx$$
$$= u + v - w$$

278

But this is not true for all values of the constants appended to each integral, but only for such as make the total constant on the second side equal to the constant on the first side.

2. $\int bu\,dx = b\int u\,dx$, b being independent of x. For differentiation gives bu for both.

3. Since $\dfrac{d}{dx}(uv) = u\dfrac{dv}{dx} + v\dfrac{du}{dx}$, the integration of both sides gives

$$\int \frac{d(uv)}{dx}dx = \int u\frac{dv}{dx}dx + \int v\frac{du}{dx}dx;$$

or (page 103.) $\quad uv = \int u\,dv + \int v\,du \qquad \int u\,dv = uv - \int v\,du$

We have thus the following theorem $\int u\,dv$ can be found whenever $\int v\,du$ can be found. The process is called *integrating by parts*, and is of fundamental importance, as we shall find.

A warning about the "arbitrary" constant

De Morgan could not recommend the shedding of the constant of integration from integration formulas without adding a word of caution [89, p. 106]:

"It must always be observed, that the arbitrary constant must never be neglected, except in finding forms, and must be applied whenever we wish to compare forms; otherwise, an integral obtained by two different methods may give two different results, *apparently*, but which, in reality, differ only by a constant. For instance, we have found by observing differentiation,

$$\int \frac{dx}{\sqrt{1-x^2}} = \sin^{-1} x \qquad \int -\frac{dx}{\sqrt{1-x^2}} = -\cos^{-1} x$$

But

$$\int -\frac{dx}{\sqrt{1-x^2}} = \int (-1)\frac{dx}{\sqrt{1-x^2}} = -\int \frac{dx}{\sqrt{1-x^2}} = -\sin^{-1} x;$$

apparently then $\cos^{-1} x = -\sin^{-1} x$, which is not true. But for the first take $\cos^{-1} x + C$, and for the second $-\sin^{-1} x + C'$, and equate these, which gives $\cos^{-1} x + \sin^{-1} x = C' - C$. But $\cos^{-1} x + \sin^{-1} x = \frac{1}{2}\pi$ a constant (p. 60); hence this comparison produces nothing except the condition that the two *constants* of *integration* here introduced must differ by $\frac{1}{2}\pi$."

Shedding the constant of integration from integration formulas saves a lot of unnecessary clutter since we will be looking, nearly always, for the functional form (the variable part) of an anti^D. We must remember, however, that the simplicity comes at a price: the proof of an identity involving integrals proves only the equality of the variable parts on the two sides. The exclusion of the constant of integration is also open to the objection that that the integral $\int(\cdots)\,dx$ now represents a particular integral, not the general indefinite integral. The former identification is almost the rule rather than the exception in books dealing with differential equations, as will become clear when we consider a few examples.

Some examples

There was a time, when nearly every fairy tale began with the words "Once upon a time, there was ...". Each of those stories could have been given a different opening, but it is nearly impossible to begin an elementary presentation of differential equations without writing down an equation of the form

$$\frac{dy}{dx} = f(x), \tag{C.1}$$

where, for the sake of keeping the discussion general, $f(x)$ is meant to be some abstract, unspecified function. The authors who make such a start know that, in any other context, the notation

$$\int f(x)\,dx \tag{C.2}$$

will be understood by the reader to signify an infinite family (of functions) each member of which, when differentiated with respect to x, yields $f(x)$. Still, the next equation in the story is stated as

$$y = \int f(x)\,dx + C, \tag{C.3}$$

280

and C is said to be an arbitrary constant. It is clear that the integral on the right-hand side is now a particular integral, not the indefinite integral representing the above mentioned family.

Why don't these authors introduce a particular antiderivative F by the relation $D_x F = f$, and write

$$y = F(x) + C \qquad\qquad\qquad (C.4)$$

instead of Eq. (C.3)? The answer is that they need, on the right-hand side, an arbitrary constant *and*—sooner or later—an integral. The only way out of the impasse is to use an existing symbol that distinguishes between the indefinite integral and a particular integral or invent a new one that respects the distinction. Let us first look at some examples of the attitude described above.

Boole [199, p. 2]: In the Integral Calculus, on the other hand, it is the relation among the primitive variables, x, y, &c. which is *sought*. In that branch of the Integral Calculus with which the student is supposed to be already familiar, the differential coefficient $y\frac{dy}{dx}$ already being given in terms of the independent variable x; it is proposed to determine the most general relation between y and x. Expressing the given relation in the form

$$\frac{dy}{dx} = \phi(x). \qquad\qquad (\text{Boole-1})$$

the relation sought is exhibited in the form

$$y = \int \phi(x)\,dx + c,$$

where the symbol $\int$ denotes a certain process of integration, the study of the various forms and conditions of which is, in a peculiar sense, the object of this part of the Integral Calculus.

Bateman [200, p. 2]: A differential equation generally possesses an infinite number of solutions, as might be inferred from an examination of a simple equation,

$$\frac{dx}{dt} = f(t)$$

which is satisfied by a relation of type

$$x = \int f(t)\, dt + c$$

where c is an arbitrary constant. An equation will be regarded as solved when an expression is obtained for the most general function which satisfies it, even if this expression involves integrals which cannot be evaluated by the ordinary methods of the integral calculus.

Wylie [201, pp. 1–2]: ...there are several large and very important classes of equations for which solutions can readily be found. For instance an equation such as

$$\frac{dy}{dx} = f(x)$$

is really a differential equation and

$$y = \int f(x)\, dx + c$$

even when the integral cannot be evaluated in terms of elementary functions, is a solution for any constant c.

Braun [202, p. 2]: If we assume that our mathematical background consists of just elementary calculus then the very sad fact is that the only first-order differential equation we can solve at present is

$$\frac{dy}{dt} = g(t), \qquad\qquad \text{(Braun-1)}$$

where g is any integrable function of time. To solve Equation (2) simply integrate both sides with respect to t, which yields

$$y(t) = \int g(t)\, dt + c.$$

Here c is an arbitrary constant of integration, and by $\int g(t)dt$ we mean an anti-derivative of g, that is, a function whose derivative is g.

Reuter [203, pp. 1–2]: One [differential equation] will certainly be known to the reader: the equation

$$\frac{dy}{dx} = f(x). \qquad\qquad \text{(Reuter-2)}$$

Here we have to find a function y whose derivative equals the given function f, and this is simply the problem of integrating f. Thus y must be an *indefinite integral* of f:

$$y = \int f(x)\,dx + C, \qquad \text{(Reuter-3)}$$

where C can be any constant. The general solution of (2) therefore involves one *arbitrary constant* whose value is at our disposal. In practice it is often not the general solution which is needed, but ...Generally, if an *initial condition*:

$$y = y_0 \quad \text{when} \quad x = x_0,$$

is imposed then the value of the arbitrary constant C in (3) can be determined. The required solution is then most simply written in terms of a *definite integral* of f, as

$$y = y_0 + \int_{x_0}^{x} f(\xi)\,d\xi. \qquad \text{(Reuter-4)}$$

A different expression for the general solution

It follows from what has been said in Chapter 3 that the most general function $G(x)$ whose derivative is equal to $f(x)$ (which is the same thing as the general solution of Eq. (C.1) must be of the form

$$G(x) = \int_{a}^{x} f(u)\,du + C, \qquad \text{(C.5)}$$

where C is some constant.

In order to elaborate on the above, a long passage by Child will now be quoted [204]:

Indefinite integral introduced as a definite integral in which the upper limit is x, and the lower an arbitrary constant A, or, preferably the lower limit omitted and supposed to be some constant, real or imagined, which makes the integral function zero.

In this way the beginner is from the first convinced that the integral is a function of the limits and that, say,

$$\int^{x} x^n\,dx = \int^{x} u^n\,du = \int^{x} v^n\,dv = x^n/(n+1) - K,$$

where K is some constant depending on the bottom arbitrary constant limit, and therefore itself arbitrary; and that K is supposed to be zero when the integral is taken as an indefinite integral, which is then considered as a function of the top limit only.

According to the definition given by Child, the notation

$$\int^x f(x)\,dx$$

denotes any antiderivative of $f(x)$ to which no constant has been purposely added. If we adopt this definition, Eq. (C.5) can be expressed as

$$G(x) = \int^x f(u)\,du + C, \tag{C.6}$$

and we will now look at a few authors who follow this convention.

Hildebrand [103, p. 1]: A *differential equation* is an equation relating two or more variables in terms of derivatives or differentials. Thus, the simplest differential equation is of the form

$$\frac{dy}{dx} = f(x) \tag{Hildebrand-1}$$

where $f(x)$ is a given function of the independent variable x. The solution is obtained immediately by integration, in the form

$$y = \int^x f(x)\,dx + C \tag{Hildebrand-2}$$

where a convenient lower limit is assumed in the integral, and C is an arbitrary constant. Whether or not it happens that the integral can be expressed in terms of simple functions is incidental, in the sense that we define a solution of a differential equation to be any functional relation, not involving derivatives or integrals of unknown functions, which implies the differential equation.

Bender & Orszag (B&O) [205, p. 3]: The most general separable equation is

$$y'(x) = a(x)b(y) \tag{B&O-1.1.2}$$

Direct integration gives the general solution

$$\int^y \frac{dt}{b(t)} = \int^x a(s)\,ds + c_1, \tag{B&O-1.1.3}$$

where c_1 is a constant of integration. [The notation $\int^x a(s)\,ds$ stands for the antiderivative of $a(x)$.]

Arfken [206–208]: This notation has been used throughout in books (co)authored by Arfken. To provide a sepcific page reference, we mention the first two equations on p. 337 of the third book [208], where the result of integration of the equation

$$\frac{dy}{y} = -p(x)\,dx$$

is stated as $\qquad \ln y = -\int^x p(x)\,dx + C.$

Forsyth [209, 210]. Forsyth's treatise went through six editions. On most occasions, he follows the authors quoted above, and denotes a particular integral by $\int f(x)\,dx$. For example, after integrating the equation [209, §50, p. 72] and [210, §50, p. 87]

$$\frac{d^n y}{dx^n} = X$$

he writes: $\dfrac{d^{n-1}y}{dx^{n-1}} = \displaystyle\int X\,dx + A_1$, A_1 being an arbitrary constant.

On other occasions, he uses the x-superscripted to denote a particular integral. For example, on he integrates [209, §66, p. 102] and [210, §66, p. 116]

$$\frac{dA}{dx} = xe^{-x},$$

and writes the result as

$$A = E + \int^x \xi e^{-\xi}\,d\xi.$$

where E is the constant of integration.

Since both

$$\int^x f(x)\,dx \quad \text{and} \quad \int^\dagger f(x)\,dx,$$

denote an antiderivative of $f(x)$ to which no constant has been purposely added, a reader may demand the justification for the introduction of a new symbol. To such a reader I will point out that the older symbol is has been used also for other purposes.

Franklin [100, p. 202]: Any function having $f(x)$ as its derivative is called an indefinite integral of $f(x)$. It is clear that if $F(x)$ is an indefinite integral of $f(x)$, then $F(x)+c$ is also an indefinite integral of $f(x)$, and we have just proved that all indefinite integrals are of this form. The symbol $\int f(x)\,dx$ or $\int^x f(x)\,dx$ is used to mean any indefinite integral.

Cohen [211, pp. 9–10]: He considers two equations,

$$\frac{\partial u}{\partial x} = M, \quad \text{and} \quad \frac{\partial u}{\partial y} = N, \qquad \text{(Cohen:3)}$$

in which M and N are functions of x and y, integrates the first and writes the result as

$$u = \int^x M\,dx + Y(y) \qquad \text{(C.7)}$$

and a footnote explains the notation: "By $\int^x M\,dx$ we mean the result of integrating $M\,dx$ considering y as a constant."

f in f_i, the expressions of f and f_i in f_2, the expressions of f, f_i and f_2 in f_3, and so on, so $f, f_i,$ cdots, f_n all have functions of $x, x_i,$ cdots, x_n and the determinant of these functions According to the theorem proved in the previous paragraph, it is equal to the product If we now set the determinant for this product in the transformed multiple integral, we obtain

Appendix D

Quadling's Quandary

Douglas A. Quadling (1926–2015) was a highly respected mathematics teacher and author of several textbooks, besides being one of the four "founding fathers" of the School Mathematics Project, which radically changed the course of mathematics teaching in Britain [212]. In an article published in the *Mathematical Gazette* [90], he raised a number of questions which deserve to be widely discussed. Using integration by parts to illustrate the difficulties that confront students (and their teachers, one might add), when "they find indefinite integrals by a method which brings them back to the integral they started with", including the paradoxical result $J = 1 + J$, where $J = \int [f'(x)/f(x)]dx$. Quadling concluded his articles with a few questions, among them the following: What *is* the best way to explain indefinite integrals to students? Ought the rule for integration by parts be stated with or without an arbitrary constant on the right-hand side? The purpose of this appendix is to answer the questions which troubled Quadling.

Quadling's questions

The first numbered equation in Quadling's note,

$$\int u \frac{dv}{dx} \, dx = uv - \int v \frac{du}{dx} \, dx, \tag{Q1}$$

embodies "the rule for integration by parts, given in most books (and in the lists of formulae provided by examination boards)".

In what follows, equation (n) of Quadling will be labelled as (Qn), and unnumbered equations following (Qn) as (Qna), (Qnb), etc.

Quadling applies Eq. (Q1), taking $u = \sin x$ and $v = \cos x$, to arrive at the result

$$\int \sin x \, \cos x \, dx = \sin x \, \sin x - \int \sin x \, \cos x \, dx. \tag{Q4}$$

Before considering the next integral, he presents a fallacy [213, 214], but this will be discussed in the next section. For integrating $e^{2x} \sin x$, he applies Eq. (Q4) twice—first with $u = e^{2x}$, $v = -\cos x$, and next with $u = 2e^{2x}$, $v = \sin x$—to find:

$$\int e^{2x} \sin x \, dx = -e^{2x} \cos x + 2e^{2x} \sin x - 4 \int e^{2x} \sin x \, dx. \tag{Q5}$$

He goes on to remark: "But the student who uses these examples to conclude that

$$\int \sin x \, \cos x \, dx = \tfrac{1}{2} \sin^2 x, \tag{Q5a}$$

$$\int e^{2x} \sin x \, dx = \tfrac{1}{5} e^{2x} (2 \sin x - \cos x) \tag{Q5b}$$

will lose a mark for 'omitting the arbitrary constant'. Where should it have appeared?".

At this point Quadling expresses some qualms about a statement that had been made, in a book co-authored by him, after the addition of an arbitrary constant to the right-hand of Eq. (Q5b):

> The arbitrary constant seems to be rather an afterthought! When you integrate by parts, it is usually only necessary to include a constant when you *carry out the final integration*; but in this example you *never actually do the final integration*. But you must not forget the constant. (My italics.)

Quadling cites Maxwell [215] for a fallacy: taking $u = \dfrac{1}{x}$ and $v = x$, he finds

$$\int \frac{1}{x}\,dx = \int \frac{1}{x} \cdot 1 \,dx = \frac{1}{x}\,x - \int x\left(-\frac{1}{x^2}\right)dx = 1 + \int \frac{1}{x}\,dx$$
(Q3)

Quadling concludes: 'Hence 0=1!'.

I will follow Quadling and call it Maxwell's fallacy, although it can be traced at least as far back as 1927 [213]. According to Quadling, the fallacy lies in interpreting the symbol $\int dx/x$ as if it represented the same primitive of $1/x$ on both sides of Eq. (Q3).

. . . .

The root of the problem therefore lies in the interpretation placed on the indefinite integrals in the statement of the rule for integration by parts in (1). [If each side stands for an infinite set of primitives], then the two sets are equal as stated. But if the symbols $\int u\dfrac{dv}{dx}dx$ and $\int v\dfrac{du}{dx}dx$ are taken to represent particular primitives of the integrands, then we can only assert that

$$\int u\frac{dv}{dx}\,dx = uv - \int v\frac{du}{dx}\,dx + c, \text{ where } c \in \mathbb{R},$$
(Q6)

This presents no difficulty in elementary applications of the method such as (2), since the arbitrary constant arises naturally in finding the indefinite integral on the right. But in proceeding to a conclusion from (4), we assumed that the symbol $\int \sin x \cos x\,dx$ represents the same primitive of $\sin x \cos x$ at both appearances. This is only valid if integration by parts is applied in the form

$$\int \sin x \, \cos x \, dx = \sin x \, \sin x - \int \sin x \, \cos x \, dx + c.$$
(Q6a)

$$\therefore \int \sin x \, \cos x \, dx = \tfrac{1}{2}\sin^2 x + \tfrac{1}{2}c$$
(D.1)

I have quoted enough, but I cannot begin without recalling Quadling's questions; to this end, I must quote his final paragraph:

So what is the best way to explain indefinite integrals to A-level students? Is it worth introducing the 'set of functions' interpretation, or should we stick with the 'general integral function' of the textbook statements? *Ought the rule for integration by parts be given in the form (1) or (6)?* What has worked best for your students? (My italics.)

Since the first two questions have already been answered in the earlier chapters of this book, it will be enough for us to focus on the italicized question; to this end, we will examine the steps involved in the process known as integration by parts in meticulous detail, which would not have been justified in the absence of Quadling's paper. To be specific, whenever an integration is carried out, a constant of integration will be slavishly added.

Integration by parts

Let us introduce the following abbreviations:

$$f_1 = u\frac{dv}{dx}, \quad I_1 = \int f_1\,dx, \quad f_2 = v\frac{du}{dx}, \quad I_2 = \int f_2\,dx, \qquad \text{(D.2)}$$

As the phrase *integration by parts* implies, a prerequisite to using the trick is to split the given integrand into two factors (or "parts"), and identifying one part with u and the other with dv/dx; one then differentiates the first part to get du/dx and integrates the second to find an anti^D:

$$\int \frac{dv}{dx}\,dx = v + \alpha \qquad \text{(D.3)}$$

Multiplying $v + \alpha$ by u, differentiating the product with respect to x, and using $d(uv)/dx = u(dv/dx) + v(du/dx)$, we get

$$\frac{d}{dx}[u(v + \alpha)] = u\frac{dv}{dx} + v\frac{du}{dx} + \alpha\frac{du}{dx}. \qquad \text{(D.4)}$$

We now integrate both sides of Eq. (D.4) with respect to x, and write the result of integration as

$$uv + \alpha u + \beta = I_1 + I_2 + \alpha u + \gamma, \qquad \text{(D.5)}$$

where β and γ are constants of integration.

Since the two terms containing α in Eq. (D.5) cancel each other, one may set $\alpha = 0$, as a result of which the arbitrary constant γ, which arises upon integration of the term $\alpha(du/dx)$, would also disappear. Therefore, Eq. (D.5) can be replaced by

$$I_1 = uv - I_2 + \beta, \qquad (D.6)$$

which is a compact form of Eq. (Q6).

Normally, the constant β is also dropped, because evaluation of the indefinite integral I_2 gives rise to a constant of integration anyway, and this reasoning leads one to Eq. (Q1). "What," the reader would surely ask, "is an *abnormal* situation?"

Three abnormal situations will now be considered.

First abnormality. This arises when

$$u\frac{dv}{dx} = v\frac{du}{dx}, \quad \text{that is,} \quad f_1 = f_2, \qquad (D.7)$$

in which case I_1 and I_2 can differ at most by a constant, but no loss of generality is incurred by ignoring this constant and lumping it with the arbitrary constant c. We can therefore write $2I_1 = uv + \beta$, or

$$I_1 = \tfrac{1}{2}uv + c. \qquad (D.8)$$

Equation (D.7) will be satisfied when $u^2 = v^2$. We apply Eq. (D.8) to one of the integrals considered by Quadling:

$$\int \sin x \cos x \, dx = \begin{cases} \tfrac{1}{2}\sin^2 x + c_1, & u = \sin x = v; \qquad (D.9) \\ -\tfrac{1}{2}\cos^2 x + c_2, & u = \cos x = -v. \quad (D.10) \end{cases}$$

Let us introduce the symbols

$$J_s = \int e^{ax} \sin bx \, dx, \quad J_c = \int e^{ax} \cos bx \, dx. \qquad (D.11)$$

The second integral considered by Quadling is a special case of J_s with $a = 2$ and $b = 1$. Making the identification $u = \sin bx$ and $dv/dx = e^{ax}$, so that $du/dx = b \cos bx$ and

$$v = \int e^{ax} \, dx = \frac{e^{ax}}{a} + A_1.$$

We have seen that inclusion of the constant A_1 is unnecessary.

292

and apply integration by parts to J_s and subsequently to J_c:

$$J_s = \sin bx \frac{e^{ax}}{a} + \beta_1 - \frac{b}{a} J_c \qquad (D.12)$$

$$= \sin bx \frac{e^{ax}}{a} + \beta_1 - \frac{b}{a} \left[\cos bx \frac{e^{ax}}{a} + \beta_2 + \frac{b}{a} J_s \right] \qquad (D.13)$$

$$\therefore \left(1 + \frac{b^2}{a^2}\right) J_s = \frac{e^{ax}(a \sin bx - b \cos bx)}{a^2} + \beta_1 - \beta_2 \qquad (D.14)$$

and $$J_s = \frac{e^{ax}(a \sin bx - b \cos bx)}{a^2 + b^2} + c \qquad (D.15)$$

Second abnormality. Another abnormal situation is that when $uv = P$ (any constant whatsoever), $d(uv)/dx = 0$, and

$$u\frac{dv}{dx} = -v\frac{du}{dx}, \quad \text{or} \quad f_1 = -f_2.$$

The above result can also be obtained by using the relations $u = P/v$ and $v = P/u$, and taking the steps shown below (which help camouflage the fallacy):

$$u\frac{dv}{dx} = \frac{P}{v} \cdot \frac{d}{dx}\frac{P}{u} = \frac{P}{v} \cdot \frac{-P}{u^2} \cdot \frac{du}{dx} = -v\frac{du}{dx}. \qquad (D.16)$$

We are now ready to dissect Maxwell's fallacy, which leads to the conclusion $J = 1 + J$, where $J = \int dx/x$. The argument starts with Eq. (Q1), which will now be rephrased as $I_1 = P - I_2$, and the paradox is fabricated by taking the following steps:

$$I_1 = P - I_2 \qquad \text{(Step 1)}$$
$$= P - (-I_1) \qquad \text{(Step 2)}$$
$$= P + I_1 \qquad \text{(Step 3)}$$
$$\therefore 0 = 1. \qquad (D.17)$$

Now, $f_1 = -f_2$ and $I_1 = \int f_1\,dx$, but we must resist the temptation to take Step 2, where $I_2 = \int f_2\,dx = \int(-f_1)\,dx$, was replaced by $-I_1$, for that contradicts $I_1 = P - I_2$, the formula with which one starts integration by parts.

All other demonstrations of this paradox [216–224], arising from different choices for u and v, entail the same error. The claim that the inclusion of an arbitrary constant ($c = -uv$) would dissolve the paradox is not valid, because such a constant cannot be called arbitrary. This point was well made in a comment entitled "How arbitrary can a constant be?" [217], the text of which (together with an adjoining comment by the editor, who happened to be none other than Quadling) is reproduced below:

Integration by parts gives

$$\int \tan x \, dx = \int \sec x . \sin x \, dx$$
$$= \sec x \cdot (-\cos x) - \int \sec x \tan x \cdot (-\cos x) \, dx + c,$$

where c is an arbitrary constant. Hence

$$\int \tan x \, dx = -1 + \int \tan x \, dx + c$$

and thus $c = 1$, contrary to the fact that it is arbitrary."

[I first heard of this pleasant fallacy through ..., who had it from a student at .... Being reminded of it by Dr. Griffiths, it seemed worth drawing to the attention of readers. It makes an admirable introduction to a discussion on "What is an indefinite integral?". D.A.Q.]

Lst us note for the sake of completeness that a consistent use of the integration formula $I_1 = P - I_2$ goes along a different route, and produces a tautology:

$$I_1 = P - I_2$$
$$= P - [P - I_1]$$
$$= I_1.$$

Third abnormality. The last abnormal situation to be considered here arises when $uv = w + K$, K being a constant. An item, contributed by Robert Weinstock [219], involved repeated (twice) integration by parts, first of $W_1 = \int e^x \sinh x \, dx$, and next of $W_2 = \int e^x \cosh x \, dx$. In the first round, $uv = e^x \cosh x = w + K$, where $w = \frac{1}{2} e^{2x}$ and $K = \frac{1}{2}$; in the second round $uv = e^x \sinh x = w - K$. Let us look at his argument now.

$$W_1 \equiv \int e^x \sinh x \, dx = e^x \cosh x - W_2 \qquad \text{(D.18)}$$

$$= e^x \cosh x - [e^x \sinh x - W_1] \qquad \text{(D.19)}$$

Cancelling W_1 from both sides, one gets

$$e^x(\cosh x - \sinh x) = 0 \quad \Longrightarrow \quad \cosh x = \sinh x. \qquad \text{(D.20)}$$

Weinstock offered the following refutation of the argument which leads to the paradox: "If, however, one is more careful and inserts an arbitrary constant with each integration by parts, one readily concludes that $c = e^x(\cosh x - \sinh x) = 1$ for all real c." But this amounts to asserting that c is not an arbitrary constant.

When $uv = w \pm P$, $d(uv)/dx = dw/dx$, and the first term on the right-hand side of Eq. (Q1) takes the form

$$\int u \frac{dv}{dx} dx = w - \int v \frac{dv}{dx} dx, \qquad \text{(D.21)}$$

and Eq. (D.19) now leads not to a paradox but to a bland identity (or a blind alley):

$$W_1 = w - \int e^x \cosh x \, dx = w - \left[w - \int e^x \sinh x \, dx \right] = W_1.$$

The purpose of an integration formula

Those who are aware that Bertrand Russell defined mathematics "as the subject in which we never know what we are talking about, nor whether what we are saying is true" [225] would not be taken aback if I now remark that integration by parts appears to be an exercise in which *we* (or some of us) never know when to include or leave out an arbitrary constant, nor whether the constants, when purposely added, are arbitrary or not.

Before we discuss the purpose of an integration formula, it will be well to note that, in the parlance of differential equations, $y = u^2$ is a particular integral of the inhomogeneous differential equation

$$\frac{dy}{dx} = 2u \frac{du}{dx}, \qquad \text{(D.22)}$$

and $y = u^2 + A$ is its general solution, and that A is the general solution of the corresponding homogeneous equation

$$\frac{dy}{dx} = 0. \tag{D.23}$$

When one is faced with the problem of finding the general solution to an inhomogeneous differential equation, the main task is that of finding a particular integral. Similarly, the main problem in integration is to find a particular primitive of the given integrand, the purpose of any integration formula being merely the simplification of the search, and a given formula is to be considered correct if, upon differentiation, the two sides are found to be equal; likewise, a particular primitive found by using a formula is to be regarded as correct if its derivative is identical with the given integrand. Adding an arbitrary constant, though sometimes necessary, is no hardship. Eq. (Q1) is to be interpreted as a recipe for finding a particular primitive, and should therefore be written, using the notation introduced in Chapter 2, as

$$\int^\dagger u \frac{dv}{dx}\, dx = uv - \int^\dagger v \frac{du}{dx}\, dx. \tag{D.24}$$

Even if one does not add an *arbitrary* constant, the process of seeking an antiderivative may itself lead to an additive but non-arbitrary constant.

Appendix E

Jacobi's Jewels

WRITTEN in Latin, "De Determinantibus functionalibus", a landmark paper by Carl Gustav Jacob Jacobi [226], will be intelligible to few of my readers. Fortunately, an annotated German version by P. Stäckel was published under the title "Ueber die Functionaldeterminanten" [117]. Jacobi's tract is divided into nineteen sections (numbered but not subtitled), the first of which, being of an introductory nature, is not of much interest. Sections 2 and 19 contain material that is highly relevant to the contents of this book; the purpose of this appendix is to provide an English rendering of Stäckel's translation of these sections.

By way of background, I would like to draw the reader's attention to the fact that, in Jacobi's time, most authors designated a differential (of x) by the symbol dx, but the alternatives dx and ∂x were not unknown; in fact, one can see in the same volume where Jacobi's article appeared, Stern (p. 6) and Oettinger (p. 174) writing $\frac{\partial Fx}{\partial x}$ and $\frac{\partial fx}{\partial x}$ for the x-derivative of a function (F and f) that depends on a single independent variable x, and Enke (p. 227) writing $\frac{dfx}{dx}$. One can also find the same author using ∂ for our d in 1832 [130], and for our D in 1837 [129]. Among those who used the upright form 'd' also in $\int f \, dx$, it will be enough to mention Lacroix [227].

298

Much earlier, Condorcet used both ∂z and dz to denote differentials of a given variable with respect to different independent variables [228, 229]. As far as I can tell, the symbol ∂ was also interpreted as the letter 'd', and this rounded form can be seen in many hand-written documents, especially in words ending in 'd', as may be seen in the following illustration, which shows three words in the present author's handwriting; these words have been chosen because they occur in the first paragraph of the draft of the Declaration of Independence (source: `http://www.gute nberg.org/files/16781/16781-h/images/decl.jpg`), and the entire draft has several words ending in a rounded 'd' and even *accordingly*.

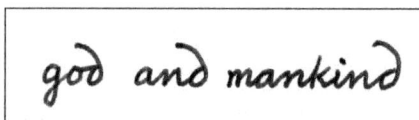

Shortly afterwards, Legendre introduced the notation $\dfrac{\partial v}{\partial x}$ for the partial derivative and $\dfrac{dv}{dx}$ for the total derivative [230, footnote on p. 8], but he made no attempt to promote the idea by repeatedly using the nuanced notation.

The freedom enjoyed by the authors in the first half of the eighteenth century and the strain placed on *their* readers by the anarchy cannot be grasped without leafing through the scientific literature of the period. Suffice it to say here that ∂ was frequently used even in one-variable (differential and integral) calculus.

In blackletter/fraktur/Gothic fonts, lower case 'd' has a similar, though more angular, shape; the output for the text *god and mankind* using three fonts available in LaTeX is shown below:

Excerpts from section 2 of Jacobi's tract

We begin at the start of the section:

> To distinguish the partial derivatives from the ordinary, where all variable quantities are regarded as functions of a single one, it has been the custom of Euler and others to enclose the partial derivatives within parentheses. As an accumulation of parentheses for reading and writing is rather tedious, I have preferred to designate by the character
>
> $$d$$
>
> ordinary differentials, and by the character
>
> $$\partial$$
>
> partial differentials. Adopting this convention, error is eliminated. If we have a function f of x and y, I shall write accordingly
>
> $$df = \frac{\partial f}{\partial x}\, dx + \frac{\partial f}{\partial y}\, dy.$$
>
> If a function contains only a single variable, one may use the character d or ∂ indifferently. The same distinction may also be made in the representation of integrations, so that the expressions
>
> $$\int f(x,y)\, dx, \quad \int f(x,y)\, \partial x$$
>
> have different meanings; in the former, y, and hence $f(x,y)$, are regarded as the function of x, and in the latter we have to carry out the integration solely with respect to x, taking y to be constant during the integration.

Jacobi goes on the discuss the relative merits of his notation and that of earlier authors, particularly Euler and Lagrange. This part of the section can be easily dispensed with.

> Let f be a function of $x, x_1, x_2, \cdots, x_n$. Assume any n functions $w, w_1, w_2, \cdots, w_n$ of these variables and think of f as a function of the variables $x, w_1, w_2, \cdots, w_n$. Then, if $x_1, x_2, \cdots, x_n$ remain constant, $w_1, w_2, \cdots, w_n$ are no longer constants when x varies, and also, when $w_1, w_2, \cdots, w_n$ remain constant, $x_1, x_2, \cdots, x_n$ do not remain constant. The expression $\frac{\partial f}{\partial x}$ will signify entirely different values according as one or the other magnitudes are held constant during the differentiation.

For example, let us introduce in a function f of the two variables x and y, instead of y as the second independent variable, an arbitrary function u of x and y. Then the differential that was earlier

$$\frac{\partial f}{\partial x},$$

now becomes

$$\frac{\partial f}{\partial x} + \frac{\partial f}{\partial u} \cdot \frac{\partial u}{\partial x}$$

Therefore, whenever partial differentials are used in this and other tracts, I shall not only imply that f depends on these variables, that it is a function of the variables $x, x_1, \ldots, x_n$ that it remains constant if they remain constant, and change if they change; and that would hold in the same way, if instead of $x, x_1, \ldots, x_n$ any other variable $w, w_1, \ldots, w_n$ functions of it were introduced as independent variables. If I say that f is a function of the variables $x, x_1, \ldots, x_n$, I wish it to be understood that, if this function is differentiated partially, the differentiation shall be so carried out that always only one of these varies, while all the others remain constant.

By the same token the formulas are to be completely unambiguous, then the notation should indicate not only the variable with respect to which the differentiation takes place but also the entire system of independent variables whose function is to be partially differentiated, so that by the notation itself one may recognize also the variables which remain constant during the differentiation. This is all the more necessary, for it is not possible to avoid the occurrence, in the same calculation and even in one and the same formula, of partial derivatives which refer to different systems of independent variables, as, for example, in the expression stated above

$$\frac{\partial f}{\partial x} + \frac{\partial f}{\partial u} \cdot \frac{\partial u}{\partial x},$$

in which f is taken as a function of x and u, whereas u is a function of x and y. The $\dfrac{\partial f}{\partial x}$ changed into this expression when u was introduced in the place of y as an independent variable. If however we write down, besides the dependent variable, also the independent variables occurring in the partial differentiations, then the above expression can be represented by the following formula which is free of every ambiguity:

$$\frac{\partial f(x,y)}{\partial x} = \frac{\partial f(x,u)}{\partial x} + \frac{\partial f(x,u)}{\partial u} \cdot \frac{\partial u(x,y)}{\partial x}.$$

P. Stäckel added an editorial note (Number 5) at this point [117, p. 65]: "This notation is ambiguous too, for the symbol is used in two different meanings, in as much as $f(x,y)$ is not the same function of x and y as $f(x,u)$ is of x and u."

Section 19 of Jacobi's tract

The theorem that I have proved in the preceding paragraph is based on the general formulas that are used to transform multiple integrals.

Consider the multiple integral

$$\int U \, \partial f \, \partial f_1 \cdots \partial f_n,$$

where U denotes a given function of $f, f_1, \cdots, f_n$. In the most general way, integration is carried out by sequentially carrying out the integration according to one variable, while the other variables are regarded as constants, in such a way that the limits of integration are functions of the other variables. Integrating first into f_n, the bounds are functions of $f, f_1 \cdots, f_{n-1}$ integral. The integral so found is again integrated with respect to f_{n-1}, and the limits of integration are functions of $f, f_1, \cdots, f_{n-1}$ and so on, until all integrations are performed.

For these multiple integrals there is a theorem which must be considered as a principle in this theory, namely that if the function U within the limits of integration never becomes infinite, the order of the integrations can be changed in any way, so that the choice of the variable with respect to which the first, or the second, or a subsequent integration takes place is immaterial provided only that the limits of the new integrations are correctly determined. This principle is self-evident if the value of the multiple integral is defined as the limit of a finite sum at steadily decreasing intervals.

With the help of this principle an immediate answer can be given to the question as to which expression under the sign of multiple integration for the element

$$\partial f \, \partial f_1 \cdots \partial f_n$$

is to be used if instead of the variables $f, f_1, \cdots, f_n$ other variables are introduced.

The first integration would be carried out with respect to f_n. For this variable, a substitute x_n is introduced by stating that f_n

is some function of x_n and the remaining members of the old set $f, f_1, \cdots f_{n-1}$, which are to be regarded as constant. If, instead of the ∂f_n differential, the integration is with respect to x_n, we have the equivalent expression

$$\partial f_n = \left(\frac{\partial f_n}{\partial x_n}\right) \partial x_n$$

so that the multiple integral presented is equal to the following:

$$\int u \left(\frac{\partial f_n}{\partial x_n}\right) \partial f \, \partial f_1 \, \cdots \, \partial f_{n-1} \, \partial x_n,$$

Let us now change the order of integrations and do the first integration not with respect to x_n, but to f_{n-1}. Instead of f_{n-1}, we introduce another variable x_{n-1} by stating that f_{n-1} is some function of x_{n-1} and it may also depend on the remaining set $f, f_1, \cdots f_{n-2}, x_n$, which are considered as constants in this first round of integration. Then the presented integral

$$\int u \left(\frac{\partial f_n}{\partial x_n}\right) \partial f \, \partial f_1 \, \cdots \, \partial f_{n-2} \, \partial x_n \, \partial f_{n-1}$$

$$= \int u \left(\frac{\partial f_{n-1}}{\partial x_{n-1}}\right) \left(\frac{\partial f_n}{\partial x_n}\right) \partial f \, \partial f_1 \, \cdots \, \partial f_{n-2} \, \partial x_n \, \partial x_{n-1}.$$

Again, the order of integrations is changed, and the first integration is made not with respect to x_{n-1} but with respect to f_{n-2}, for which a new variable x_{n-2} is introduced. By always changing the order of the integrations after the introduction of a new variable, and introducing a variable instead of the variable with respect to which the first integration is to be carried out, one finally arrives at the following statement for the transformed integral:

$$\int u \left(\frac{\partial f}{\partial x}\right) \left(\frac{\partial f_1}{\partial x_1}\right) \cdots \left(\frac{\partial f_n}{\partial x_n}\right) \partial x \, \partial x_1 \, \cdots \, \partial x_n.$$

In the transformed expression f_n is a function of $f, f_1, \cdots, f_{n-1}$, x_n; f_{n-1} is a function of $f, f_1, \cdots, f_{n-2}, x_{n-1}, x_n$, and generally f_i a function of $f, f_1, \cdots, f_{i-1}, x_i, x_{i+1}, x_n$, so that the last function f contains all new variables $x, x_1, \cdots, x_n$. But if one sets the expression f in f_1, the expressions of f and f_1 in f_2, the expressions of f, f_1 and f_2 in f_3, and so on, so that $f, f_1, \cdots, f_n$ are all functions of $x, x_1, \cdots, x_n$ and the determinant of these functions

$$\sum \pm \frac{\partial f}{\partial x} \cdot \frac{\partial f_1}{\partial x_1} \cdots \frac{\partial f_n}{\partial x_n}$$

equals, according to the theorem proved in the previous paragraph, the product

$$\left(\frac{\partial f}{\partial x}\right) \cdot \left(\frac{\partial f_1}{\partial x_1}\right) \cdots \left(\frac{\partial f_n}{\partial x_n}\right).$$

If we now insert the determinant for this product in the transformed multiple integral, we obtain

$$\int U \, \partial f \, \partial f_1 \cdots \partial f_n = \int U \cdot \left(\sum \pm \frac{\partial f}{\partial x} \cdot \frac{\partial f_1}{\partial x_1} \cdots \frac{\partial f_n}{\partial x_n}\right) \partial x \, \partial x_1 \cdots \partial x_n.$$

and that is the general formula for the transformation of a multiple integral.

Euler and Lagrange found this formula for two and three variables at about the same time, but Euler was a little earlier.

This formula, too, manifests in an excellent manner the analogy between the differential and the functional determinant.

Two remarks are worth making on the concluding section of Jacobi's tract. First, he uses the same symbol U after each change of variable, and secondly, he mentions neither the sign of the derivative at each stage (presumably because he expects the reader to reverse the limits, nor the desirability of using the absolute of the functional determinant (now called the Jacobian determinant).

Bibliography

[1] Mark Twain (Samuel L. Clemens). *The Adventures of Huckleberry Finn*. Chatto & Windus, London, 1884. *N. B.* The first *numbered* page is p. x; the 'Notice' is on p. vii and the 'Explanatory' on p. viii).

[2] Mark Twain. *Adventures of Huckleberry Finn*. Charles L. Webster and Company, New York, 1885. *N. B.* **All** pages numbering is in Arabic numerals. The first *numbered* page is p. 10; the 'Notice' is on p. 7 and the 'Explanatory' on p. 8.

[3] Ezra Pound. *ABC of Reading*. Faber and Faber, London, 1973.

[4] E. G. Phillips. The teaching of differentials. *The Mathematical Gazette*, 15(214):401–403, 1931.

[5] J. R. Wilton, J. T. Combridge, E. G. Phillips, and T. A. A. B[roadbent]. The teaching of differentials. *The Mathematical Gazette*, 16(217):5–10, 1932.

[6] J. T. Combridge. Differentials. *The Mathematical Gazette*, 18(228):68–79, 1934.

[7] D. K. Picken. On differentials. *The Mathematical Gazette*, 19(233):79–86, 1935.

[8] G. Temple. The rehabilitation of differentials [and discussion]. *The Mathematical Gazette*, 20(238):120–131, 1936.

[9] D. K. Picken. 1208. On differentials. *The Mathematical Gazette*, 20(240):276–279, 1936.

[10] A. Robson and C. V. Durell. 1174. On differentials. *The Mathematical Gazette*, 20(237):52–53, 1936.

306

[11] R. Cooper. 1207. On differentials. *The Mathematical Gazette*, 20(240):276–276, 1936.

[12] A. Barton. Differentials from a new viewpoint. *The Mathematical Gazette*, 29(287):193–199, 1945.

[13] E. G. Phillips. On differentials. *The Mathematical Gazette*, 33(305):202–204, 1949.

[14] J. Hadamard. 2158. On differentials. *The Mathematical Gazette*, 34(309):210–210, 1950.

[15] P. Gant. 2203. Differentials. *The Mathematical Gazette*, 35(312):111–114, 1951.

[16] Hugh Thurston. A paradox. *The Mathematical Gazette*, 48(363):27–29, 1964.

[17] W. G. L. Sutton. Correspondence. *The Mathematical Gazette*, 48(366):442–444, 1964.

[18] D. A. T. Wallace. Correspondence. *The Mathematical Gazette*, 49(368):205–207, 1965.

[19] Mark Kac and J. F. Randolph. Differentials. *The American Mathematical Monthly*, 49(2):110–112, 1942.

[20] Dunham Jackson. A comment on "Differentials". *The American Mathematical Monthly*, 49(6):389–389, 1942.

[21] Alonzo Church. Differentials. *The American Mathematical Monthly*, 49(6):389–392, 1942.

[22] W. R. Ransom. Bringing in differentials earlier. *The American Mathematical Monthly*, 58(5):336–337, 1951.

[23] M. K. Fort. Differentials. *The American Mathematical Monthly*, 59(6):392–395, 1952.

[24] C. G. Phipps. The relation of differential and delta increments. *The American Mathematical Monthly*, 59(6):395–398, 1952.

[25] H. J. Hamilton. Toward understanding differentials. *The American Mathematical Monthly*, 59(6):398–403, 1952.

[26] C. B. Allendorfer. Editorial. *The American Mathematical Monthly*, 59(6):403–406, 1952.

[27] J. J. Sylvester. *The Laws of Verse: Or Principles of Versification Exemplified in Metrical Translations, Together with an Annotated Reprint of the Inaugural Presidential Address to the Mathematical and Physical Section of the British Association at Exeter*. Longmans, Green, and Co., London, 1870.

[28] K. Menger. *Calculus: A Modern Approach*. Illinois Institute of Technology, Chicago, 1952.

[29] K. Menger. *Calculus: A Modern Approach* (Second Edition). Illinois Institute of Technology, Chicago, 1952.

[30] Karl Menger. *Calculus: A Modern Approach* (Third Edition). Ginn and Company, Boston, 1955.

[31] E. T. Steller. 3248. deleted or depleted? *The Mathematical Gazette*, 53(386):412–413, 1969.

[32] Why is treating the derivative like a fraction an abuse of notation? `https://www.quora.com/Why-is-treating-the-der ivative-like-a-fraction-an-abuse-of-notation`. Online; accessed 25 February 2017.

[33] Can Leibniz notation be treated as a quotient? `http: //math.stackexchange.com/questions/4893/can-lei bniz-notation-be-treated-as-a-quotient`. Online; accessed 25 February 2017.

[34] Is $\frac{dy}{dx}$ not a ratio? `http://math.stackexchange.com/que stions/21199/is-frac-textrmdy-textrmdx-not-a-r atio`. Online; accessed 25 February 2017.

[35] How misleading is it to regard $\frac{dy}{dx}$ as a fraction? `http: //mathoverflow.net/questions/73492/how-misle ading-is-it-to-regard-fracdydx-as-a-fraction`. Online; accessed 25 February 2017.

[36] A really basic integration question concerning differentials. `http://math.stackexchange.com/questions /1296698/a-really-basic-integration-question-c oncerning-differentials`. Online; accessed 25 February 2017.

[37] Usage of dx in integrals. http://math.stackexchange.com /questions/1068/usage-of-dx-in-integrals. Online; accessed 25 February 2017.

[38] What is dx in integration? http://math.stackexchange.c om/questions/200393/what-is-dx-in-integration. Online; accessed 25 February 2017.

[39] Why absolute values of Jacobians in change of variables for multiple integrals but not single integrals? https: //math.stackexchange.com/questions/856654/why-a bsolute-values-of-jacobians-in-change-of-varia bles-for-multiple-integrals-b. Online; accessed 25 February 2017.

[40] J. G. Leathem. *Volume and Surface Integrals Used in Physics* (First Edition). Cambridge Tracts in Mathematics and Mathematical Physics. Cambridge University Press, Cambridge, 1905.

[41] J. G. Leathem. *Volume and Surface Integrals Used in Physics* (Second Edition). Cambridge Tracts in Mathematics and Mathematical Physics. Cambridge University Press, Cambridge, 1922.

[42] G. H. Hardy. *A Course of Pure Mathematics* (First Edition). Cambridge University Press, Cambridge, 1908.

[43] M. H. Protter. Calculus reform. *The Mathematical Intelligencer*, 12(4):6–9, 1990.

[44] Joan Ferrini-Mundy and Karen Geuther Graham. An overview of the calculus curriculum reform effort: Issues for learning, teaching, and curriculum development. *The American Mathematical Monthly*, 98(7):627–635, 1991.

[45] David Mumford. Calculus reform—for the millions. *Notices of the American Mathematical Society*, 44(5):559–563, 1997.

[46] David Klein and Jerry Rosen. Calculus reform—for the $millions. *Notices of the American Mathematical Society*, 44(10):1324–1325, 1997.

[47] Alexander Philip. *The Calendar: Its History, Structure and Improvement*. Cambridge University Press, Cambridge, 1921.

[48] Alexander Philip. *The Reform of the Calendar*. Kegan Paul, Trench, Trübner & Company Limited, London, 1914.

[49] M. J. O'Brien. Calendar reform. *Irish Astronomical Journal*, 3:80–83, 1954.

[50] E. L. Cohen. Adoption and reform of the Gregorian calendar. *Math Horizons*, 7(3):5–11, 2000.

[51] A. De Morgan. *The Differential and Integral Calculus*. R. Baldwin & Cradock, London, 1842.

[52] Alfred L. Roca, Nicholas Georgiadis, Jill Pecon-Slattery, and Stephen J. O'Brien. Genetic evidence for two species of elephant in Africa. *Science*, 293(5534):1473–1477, 2001.

[53] Carl Zimmer. *Bringing them back to life*. `https://www.nationalgeographic.com/magazine/2013/04/species-revival-bringing-back-extinct-animals/`.

[54] Thomas Jefferson (author) and H. A. Washington (editor). *The Writings of Thomas Jefferson: Being His Autobiography, Correspondence, Reports, Messages, Addresses, and Other Writings, Official and Private*. Volume 6. Taylor & Maury, Washington D. C., 1854.

[55] Felix Klein. *Elementary Mathematics from an Advanced Standpoint: Arithmetic, Algebra, Analysis*. Macmillan, London, 1932.

[56] G. B. Airy. *Mathematical Tracts on Physical Astronomy, the Figure of the Earth, Precession and Nutation, and the Calculus of Variations*. J. Deighton & Sons, Cambridge, 1826.

[57] John West (author), John Leslie, and Edward Sang (editors). *Mathematical Treatises: containing I. The theory of analytical functions. II. Spherical Trigonometry, with practical and Nautical Astronomy ... Edited (after the Author's Death) from his Mss. by the Late Sir J. Leslie ... Accompanied by a Memoir of the Life and Writings of the Author, by Edward Sang, F. R. S. E.* Oliver & Boyd, Edinburgh, 1838.

[58] William Bragg. Electrons and ether waves. *The Scientific Monthly*, 14(2):153–160, 1922.

[59] G. H. Hardy. *A Course of Pure Mathematics* (Third Edition). Cambridge University Press, Cambridge, 1921.

[60] Hugh Thurston. Should we reform the teaching of calculus? *The Mathematical Gazette*, 89(515):233–234, 2005.

[61] C. H. Edwards and D. E. Penney. *Calculus with Analytic Geometry* (pp. 268–270). Prentice Hall, NJ 07458, 1998.

[62] D. F. Gregory. *Examples of the Processes of the Differential and Integral Calculus*. J. and J. J. Deighton, Cambridge, 1841.

[63] H. B. Phillips. *Analytical Geometry and Calculus*. John Wiley & Sons, Inc., New York, 1962.

[64] W. E. Byerly. *Elements of the Differential Calculus: With Examples and Applications*. Ginn & Company, Boston, 1890.

[65] W. E. Byerly. *Elements of the Integral Calculus: With a Key to the Solution of Differential Equatons*. Ginn, Heath & Company, Boston, 1881.

[66] W. E. Byerly. *Elements of the Integral Calculus: With a Key to the Solution of Differential Equatons* (Second Edition). Ginn, Heath & Company, Boston, 1888.

[67] Jerome Hunsaker and Saunders Mac Lane. Edwin Bidwell Wilson (1879–1964): A Biographical Memoir. http://www.nasonline.org/publications/biogr aphical-memoirs/memoir-pdfs/wilson-edwin-b.pdf, 1973. Online; accessed 25 August 2016.

[68] E. B. Wilson. *Advanced Calculus*. Ginn and Company, Boston, 1912.

[69] J. Hadamard. 2158. On differentials. *The Mathematical Gazette*, 34(309):210–210, 1950.

[70] J. Jerome Keisler. Foundations of infinitesimal calculus. https://www.math.wisc.edu/ keisler/foundations.html.

[71] H. Jerome Keisler. *Elementary Calculus: An Infinitesimal Approach* (Third Edition). Dover Publications, Inc, New York, 2012.

[72] R. Courant, H. Robbins, and I. Stewart. *What is Mathematics? An Elementary Approach to Ideas and Methods*. Oxford Paperbacks. Oxford University Press, Oxford, 1996.

[73] G. H. Hardy. *A Course of Pure Mathematics* (Tenth Edition). The English Language Book Society, London, 1961.

[74] G. M. Fikhtengol'ts (author) and I. N. Sneddon (translator). *The Fundamentals of Mathematical Analysis*. Pergamon Press, Oxford, 1965 [p. 170].

[75] A. F. Bermant, I. G. Aramonovich (authors), V. M. Volosov, and I. G. Volosova (translators). *Mathematical Analysis: A Brief Course for Engineering Students*. Mir Publishers, Moscow, 1975 [p. 158].

[76] M. Hazewinkel. *Encyclopaedia of Mathematics* (Volume 3). Kluwer Academic Publishers, Dordrecht, 1989 [p. 105].

[77] Theodore Chaundy. *The Differential Calculus*. Clarendon Press, Oxford, 1935.

[78] C. H. Edwards and D. E. Penney. *Calculus with Analytic Geometry*. Prentice Hall, NJ 07458, 1998.

[79] A. N. Whitehead. *An Introduction to Mathematics*. Henry Holt and Company, New York, 1911.

[80] P. B. Medawar. *The Uniqueness of the Individual*. Basic Books, Inc., New York, 1957.

[81] E. W. Hobson. *The Theory of Functions of a Real Variable and the Theory of Fourier's Series*. Cambridge University Press, Cambridge, 1907.

[82] Theodore Chaundy. *The Differential Calculus* p. 75. Clarendon Press, Oxford, 1935.

[83] A. Barton. Differentials from a new viewpoint. *The Mathematical Gazette*, 29(287):193–199, 1945.

[84] Bertrand Russell. *The Autobiography of Bertrand Russell: 1872–1914*. George Allen and Unwin Ltd, London, 1967.

[85] V. S. Shipachev (author) and A. N. Tikhonov (editor). *Higher Mathematics*. Mir Publishers, Moscow, 1988.

[86] C. H. Edwards and David E. Penney. *Elementary Differential Equations*. Pearson Education, Upper Saddle River, NJ 074, 2008.

[87] Sir John Bland-Sutton. *Evolution and Disease*. Scribner & Welford, New York, 1890.

[88] G. B. Airy. *An Elementary Treatise on Partial Differential Equations: Designed for the Use of Students in the University*. Macmillan, London, 1866.

[89] A. De Morgan. *The Differential and Integral Calculus*. R. Baldwin & Cradock, London, 1836.

[90] Douglas Quadling. Integrals—Indefinite or Misdefined? *The Mathematical Gazette*, 91(521):303–307, 2007.

[91] P. J. Sloane. The significance of the "insignificant" constants. *The Mathematics Teacher*, 82(3):186–188, 1989.

[92] C. B. Allendoerfer. The case against calculus. *The Mathematics Teacher*, 56(7):482–485, 1963.

[93] R. Courant (author) and E. J. McShane (translator). *Differential and Integral Calculus* (Vol. 1). Blackie & Son, Ltd., London, 1961.

[94] N. S. Piskunov (author) and George Yankovsky (translator). *Differential and Integral Calculus*. Associated East-West Press Pvt. Ltd., New Delhi, 1956.

[95] Joseph Fourier. *Théorie de la Chaleur*. Firmin Didot, Paris, 1822.

[96] John F. W. Herschel. On a remarkable application of Cotes's theorem. *Philosophical Transactions of the Royal Society of London*, 103:8–26, 1813.

[97] J. B. J. Fourier (author) and Alexander Freeman (translator). *The Analytical Theory of Heat*. Cambridge University Press, Cambridge, 1878.

[98] E. Picard. *Traité d'analyse* Tome 1. Gauthier-Villars, Paris, 1891.

[99] M. R. Spiegel. *Schaum's Outline of Theory and Problems of Advanced Calculus*. McGraw-Hill Book Company, New York, 1963.

[100] P. Franklin. *A Treatise on Advanced Calculus*. Dover, New York, 1968.

[101] A. E. Taylor. *Calculus with Analytic Geometry*. Prentice-Hall, Eaglewoods Cliff, NJ, 1959.

[102] A. Jeffrey. *Mathematics for Engineers and Scientists*. Thomas Nelson and Sons Ltd, London, 1971.

[103] F. B. Hildebrand. *Advanced Calculus for Applications*. Prentice-Hall, Eaglewood Cliffs, NJ, 1962.

[104] Anonymous. Obituary notices of fellows deceased. *Proceedings of the Royal Society of London. Series A, Containing Papers of a Mathematical and Physical Character*, 111(759):i–xlviii, 1926.

[105] John Perry. Symbol for partial differnetiation. *Nature*, 66(1698):53, 1902.

[106] Thomas Muir. The theory of Jacobians in the historical order of its development up to 1841. *Proceedings of the Royal Society of Edinburgh*, 24:151–195, 1904.

[107] John Perry. Symbol for partial differnetiation. *Nature*, 66(1707):271–272, 1902.

[108] John S. Thomsen. Thermodynamic derivatives without tables. *American Journal of Physics*, 32(9):666–671, 1964.

[109] R. E. Sonntag, C. Borgnakke, and G. J. Van Wylen. *Fundamentals of Thermodynamics* (Sixth Edition). John Wiley & Sons, New York, 2003.

[110] T. K. Sherwood and C. E. Reid. *Applied Mathematics in Chemical Engineering*. McGraw-Hill Book Company, New York, 1939.

[111] M. Thorade and A. Saadat. Partial derivatives of thermodynamic state properties for dynamic simulation. *Environmental Earth Sciences*, 70(8):3497–3503, 2013.

[112] P. W. Bridgman. A complete collection of thermodynamic formulas. *Phys. Rev.*, 3:273–281, 1914.

[113] E. W. Dearden. Expansion formulae for first-order partial derivatives of thermal variables. *European Journal of Physics*, 16(2):76, 1995.

[114] F. Weinhold. Metric geometry of equilibrium thermodynamics. *The Journal of Chemical Physics*, 63(6), 1975.

[115] Charles Babbage. *Passages from the Life of a Philosopher*. Longman, Green, Longman, Roberts, & Green, London, 1864.

314

[116] S. L. Ross. *Differential Equations* (Third edition). John Wiley & Sons Inc., New York, 1984.

[117] C. G. J. Jacobi (author) and P. Stäckel (editor). *Ueber die Functional-determinanten*. W. Engelmann, Leipzig, 1896 (first published in 1841).

[118] George Gamow. *My World Line; An Informal Autobiography*. Viking Press, New York, 1970.

[119] Dam Thanh Son. *Igor Tamm and the Taylor expansion*. https://damtson.wordpress.com/2017/08/12/igor-tamm-and-the-taylor-expansion/.

[120] I. S. Sokolnikoff and R. M. Redheffer. *Mathematics of Physics and Modern Engineering*. McGraw-Hill, New York, 1958 (p. 143).

[121] George Arfken. *Mathematics of Physics and Modern Engineering* (Third Edition). Academic Press, Inc., San Diego, 1985 (p. 303).

[122] Oliver Heaviside. *Electromagnetic Theory* (Vol. 2). "The Electrician" Printing and Publishing Co. Ltd., London, 1899.

[123] E. J. Routh. *The Advanced Part of a Treatise on the Dynamics of a System of Rigid Bodies: Being Part II of a Treatise on the Whole Subject* (5th Edition). Macmillan & Co., London, 2013.

[124] G. H. Hardy. *Collected Papers of G. H. Hardy* Vol. 7. Clarendon Press, Oxford, 1979.

[125] Oliver Heaviside. *Electromagnetic Theory* (Vol. 1). "The Electrician" Printing and Publishing Co. Ltd., London, 1893.

[126] J. W. Strutt. *Scientific Papers* Volume 1 (1869–1881). Cambridge University Press, Cambridge, 1899.

[127] G. M. M. Mr. O. Heaviside's Electromagnetic Theory, Vol. II. (Book review). *Philosophical Magazine* (Fifth Series), 58(292):309–312, 1899.

[128] J. L. B. Cooper. Heaviside and the operational calculus. *The Mathematical Gazette*, 36(315):5–19, 1952.

[129] R. Lobatto. *Mémoire sur la théorie des caractéristiques employées dans lànalyse mathématique*. C. G. Sulpke, Amsterdam, 1837.

[130] R. Lobatto. Note sur l'intégration de la fonction $\frac{\partial z}{a + bz}$. *Journal für die reine und angewandte Mathematik*, 1832(9):259–260, 1832.

[131] D. F. Gregory. On the solution of linear differential equations with constant coefficients. *The Cambridge Mathematical Journal*, 1:22–32, 1839.

[132] Various Authors. *Memoirs of the Analytical Society*. The Analytical Society, Cambridge, 1813.

[133] John Toplis. On the decline of mathematical studies, and the sciences dependent upon them. *The Philosophical Magazine*, 20(77):249–284, 1805.

[134] John Playfair. *Traité de Méchanique Céleste* by P. S. La Place (Book Review). *The Edinburgh Review*, 11(22):249–284, 1808.

[135] John Playfair. *The Works of John Playfair, Esq.* A. Constable & Co., Edinburgh, 1822.

[136] J. R. Newman. *The World of Mathematics* Volume 1. George Allen and Unwin Ltd, London, 1960.

[137] S. F. (author) Lacroix. *an Elementary Treatise of the Differential and Integral Calculus by S. F. Lacroix. Translated from the French; with an Appendix and Notes.* J. Deighton and Sons, Cambridge, 1816.

[138] M. Kline. *Mathematical Thought from Ancient to Modern Times:*. Oxford University Press, New York, 1972.

[139] Anonymous (attributed to J. F. W. Herschel). Obituary notices of deceased fellows. *Proceedings of the Royal Society of London*, 9:536–543, 1857.

[140] Anonymous. Obituary. *Monthly Notices of the Royal Astronomical Society*, 19(4):105–156, 1859.

[141] G. B. Airy. On the disturbances of pendulums and balances, and on the theory of escapements [read november 26, 1826]. *Transactions of the Cambridge Philosophical Society*, 3(1):105–128, 1830.

[142] R. Robson and Walter F. Cannon. William Whewell, F. R. S. (1794–1866). *Notes and Records R. Soc. Lond.*, 19(2):168–191, 1964.

[143] William Whewell. *The Doctrine of Limits with its Applications.* J. and J. J. Deighton, Cambridge, 1838.

[144] Anonymous. Obituary. *Monthly Notices of the Royal Astronomical Society*, 32(4):112–118, 1872.

[145] H. Gwynedd Green and H. J. J. Winter. John Landen, F.R.S. (1719-1790)–Mathematician. *Isis*, 35(1):6–10, 1944.

[146] Christine Phili. John Landen: First Attempt for the Algebrization of Infinitesimal Calculus. In *Trends in the Historiography of Science. Boston Studies in the Philosophy of Science*, Vol. 151 (edited by K. Gavroglu, J. Christianidis, and E. Nicolaidis), pages 279–293. Dordrecht, Dordrecht, 1994.

[147] John Landen. *A Discourse Concerning the Residual Analysis: A New Branch of the Algebraic Art, of Very Extensive Use, Both in Pure Mathematics and Natural Philosophy.* J. Nourse, London, 1758.

[148] John Landen. *The Residual Analysis; A New Branch of the Algebraic Art, of Very Extensive Use, Both in Pure Mathematics and Natural Philosophy. Book I.* John Landen, London, 1764.

[149] Dugald Stewart, James Macintosh, John Playfair, and John Leslie. *Dissertations on the History of Metaphysical and Ethical, and of Mathematical and Physical Science.* Adam and Charles Black, Edinburgh, 1835.

[150] J. L. Lagrange. *Théorie des fonctions analytiques: contenant les principes du calcul différentiel, dégagés de toute considération d'infiniment petits, d'évanouissans, de limites et de fluxions, et réduits à l'analyse algébrique des quantités finies.* de l'Imprimerie de la République, Paris, 1797.

[151] A. A. Cournot, L. Hachette, and F. D. Frères. *Traite elementaire de la theorie des fonctions et de calcul infinitesimal: 1.* Chez L. Hachette, Paris, 1841.

[152] J. M. C. Duhamel. *Cours d'analyse de l'école polytechnique.* Bachelier, Paris, 1841.

[153] Claude Navier. *Résumé des lecons d'analyse données à l'Ecole polytechnique.* Carilian-Goeury et Vr Dalmont, Paris, 1840.

[154] F. N. M. Moigno and A. L. Cauchy. *Leçons de calcul différentiel et de calcul intégral: Calcul différentiel.* Mallet-Bachelier, Paris, 1840.

[155] S. Earnshaw. On the nature of the molecular forces which regulate the constitution of the luminiferous ether. *Transactions of the Cambridge Philosophical Society,* 7(1):97–112, 1842.

[156] Samuel Earnshaw. *On the Notation of the Differential Calculus.* J. & J. J. Deighton, Cambridge, 1832.

[157] J. R. Young. *The Elements of the Differential Calculus.* John Souter, London, 1833.

[158] J. R. Young. *The Elements of the Differential Calculus.* Carey, Lea & Blanchard, Philadelphia, 1833.

[159] W. C. Ottley. *A Treatise on the Differential Calculus.* W. P. Grant, Cambridge, 1838.

[160] Thomas Jarrett. On algebraic notation [read november 12, 1827]. *Transactions of the Cambridge Philosophical Society,* 3(1):65–104, 1830.

[161] Thomas Jarrett. *An Essay on Algebraic Development.* J. & J. J. Deightons, Cambridge, 1831.

[162] Anonymous (identified as Augustus De Morgan). Cambridge differential notation: On the notation of the differential calculus, adopted in some works lately published at cambridge. *Quarterly Journal of Education,* 7(January-April):100–110, 1834.

[163] W. H. Miller. *An Elementary Treatise on the Differential Calculus.* J. & J. J. Deighton, Cambridge, 1833.

[164] J. R. Tanner (editor). *The Historical Register of the University of Cambridge: Being a Supplement to the Calendar, with a Record of University Offices, Honours and distinctions to the Year 1910.* Cambridge University Press, Cambridge, 1917.

[165] G. B. Airy. On the figure of the earth. *Philosophical Transactions of the Royal Society of London,* 116(1/3):548–578, 1826.

[166] C. G. Knott (editor). *Life and Scientific Work of Peter Guthrie Tait.* Cambridge University Press, Cambridge, 1911.

318

[167] J. J. Observations on the notations employed in the differential and integral calculus. *Philosophical Magazine* (Third Series), 24(156):25–37, 1844.

[168] Rev. Roberet Murphy. *Elementary Principles of the Theories of Electricity, Heat, and Molecular Actions*. J. & J. Deighton, Cambridge, 1833.

[169] R. C. Archibald et al. *Benjamin Peirce (1809–1880)*. The Mathematical Association of America, Oberlin, Ohio, 1925.

[170] Benjamin Peirce. *An Elementary Treatise on Curves, Functions, and Forces* Volume 1. James Munroe and Company, Boston, 1841.

[171] Benjamin Peirce. *An Elementary Treatise on Curves, Functions, and Forces* Volume 1 (New Edition). James Munroe and Company, Boston, 1852.

[172] Benjamin Peirce. *An Elementary Treatise on Curves, Functions, and Forces* Volume 2. James Munroe and Company, Boston, 1846.

[173] Cleveland Abbe. Memoir of William Ferrel. `http://www.nasonline.org/publications/biographica l-memoirs/memoir-pdfs/ferrel-william.pdf`, 1895. Online; accessed 28 May 2018.

[174] W. M. Davis. Sketch of william ferrel. *Popular Science Monthly*, 40(March):686–695, 1892.

[175] William Ferrell. The motion of fluids and solids relative to the earth's surface. *Mathematical Monthly*, 1:140–148; 210–216; 366–373; 397–406, 1958.

[176] J. L. Coolidge. William elwood byerly—in memoriam. *Bulletin of the American Mathematical Society*, 42(5):295–298, 1936.

[177] A. D. D. Craik. Geometry, analysis, and the baptism of slaves: John West in Scotland and Jamaica. *Historia Mathematica*, 25(1):29–74, 1998.

[178] Various. *The Penny Cyclopædia of the Society for the Diffusion of Useful Knowledge* Volume 24. Charles Knight and Co., London, 1842.

[179] Karl Menger (author), Louise Golland, Brian McGuinness, and Abe Sklar (editors). *Reminiscences of the Vienna Circle and the Mathematical Colloquium*. Vienna Circle Collection. Springer, Dordrecht, 2013.

[180] Seymour Kass. Karl Menger. *Notices of the American Mathematical Society*, 43(5):558–561, 1996.

[181] Iain T. Adamson. Book review. *The Mathematical Gazette*, 39(329):245–250, 1955.

[182] T Erber. Modern calculus notation applied to physics. *European Journal of Physics*, 15(3):111–118, 1994.

[183] Philip S. Marcus. Comments on variable-free calculus. *Educational Studies in Mathematics*, 4(3):324–330, 1972.

[184] E. W. Hobson. *A Treatise on Plane Trigonometry* (Second Edition). Cambridge University Press, Cambridge, 1897.

[185] W. L. C. Sargent. 2213. on the differentiation of a function of a function. *The Mathematical Gazette*, 35(312):121–122, 1951.

[186] L. S. Holden. Providing motivation for the chain rule. *The Mathematics Teacher*, 60(8):850–855, 1967.

[187] Several authors. *The Penny Cyclopædia of the Society for the Diffusion of Useful Knowledge* Volume 23. Charles Knight & Co., London, 1842.

[188] Gary Perlman. Making mathematical notation more meaningful. *The Mathematics Teacher*, 75(6):462–466, 1982.

[189] Joseph Mazur. *Enlightening Symbols: A Short History of Mathematical Notation and Its Hidden Powers*. Princeton University Press, Princeton, 2014.

[190] J. A. Lohne. Essays on Thomas Harriot. I. Billiard Balls and Laws of Collision. II. Ballistic Parabolas. III. Survey of Harriot Scientific Writings. *Arch. Hist. Exact Sci.l*, 20(3–4):189–312, 1979.

[191] Anonymous. *Harriot*. http://mathematics.edwardworthlibrary.ie/notation/harriot/.

[192] Ian Bruce (translator and editor). *Harriot's ARTIS ANALYTICAE PRAXIS, ad aequationes algebraicae nova methodo resolvendas.* http://www.17centurymaths.com/contents/contentspraxis.htm.

[193] A. De Morgan. *A Treatise on the Calculus of Functions.* Baldwin and Cradock, London, 1836.

[194] E. Smedley, H. J. Rose, and H. J. Rose. *Encyclopædia Metropolitana* (Volume 2). Encyclopædia Metropolitana. B. Fellowes and others, 1845.

[195] G. G. Stokes and J. Larmor (editor). *Mathematical and Physical Papers* Vol. 1). Cambridge University Press, Cambridge, 1880.

[196] G. G. Stokes and J. Larmor (editor). *Memoir and Scientific Correspondence of the Late Sir George Gabriel Stokes* Vol. 1). Cambridge University Press, Cambridge, 1907.

[197] W. W. Rouse Ball. *Cambridge Papers.* Macmillan and Co., Ltd., London, 1918.

[198] T. B. Macaulay (author) and W. H. Hudson (editor). *Macaulay's Essay on Lord Clive.* George G. Harrap & Company, London, 1910.

[199] George Boole. *A Treatise on Differential Equations.* Macmillan and Co., London, 1859.

[200] H. Bateman. *Differential Equations.* Longmans, Green and Co., 1918.

[201] C. R. Wylie. *Differential Equations.* McGraw-Hill Kogakusha, Tokyo, 1979.

[202] M. Braun. *Differential Equations and Their Applications: An Introduction to Applied Mathematics* Third Edition. Springer-Verlag, New York, 1983.

[203] G. E. H. Reuter. *Elementary Differential Equations and Operators.* Routledge and Kegan Paul, London, 1962.

[204] J. M. Child. The approach to the differentiation and integration of logarithmic and exponential functions. *The Mathematical Gazette*, 13(182):111–114, 1926.

[205] C. M. Bender and S. A. Orszag. *Advanced Mathematical Methods for Scientists and Engineers I: Asymptotic Methods and Perturbation Theory.* Springer-Verlag New York, Inc., New York, 1999.

[206] George Arfken. *Mathematics of Physics and Modern Engineering.* Academic Press, New York, 1968.

[207] G. B. Arfken and H. J. Weber. *Mathematical Methods for Physicists.* Harcourt/Academic Press, Burlington, MA, 2001.

[208] G. B. Arfken, H. J. Weber, and F. E. Harris. *Mathematical Methods for Physicists* Seventh Edition. Elsevier, Amsterdam, 2013.

[209] A. R. Forsyth. *A Treatise on Differential Equations.* Macmillan and Co, London, 1885.

[210] A. R. Forsyth. *A Treatise on Differential Equations* Sixth Edition. Macmillan & Co Ltd, London, 1929.

[211] Abraham Cohen. *An Elementary Treatise on Differential Equations.* D. C. Heath & Co, Boston, 1906.

[212] Geoffrey Howson. Douglas Arthur Quadling 1926–2015. *The Mathematical Gazette,* 99:193–195, 7 2015.

[213] J. L. Walsh. A paradox resulting from integration by parts. *The American Mathematical Monthly,* 34(2):88–88, 1927.

[214] E. A. Maxwell. *Fallacies in Mathematics.* Cambridge University Press, 2006.

[215] E. A. Maxwell. *Fallacies in Mathematics.* Cambridge University Press, Cambridge, 1963.

[216] H. V. Lowry. 2090. 1 = 0. *The Mathematical Gazette,* 33(306):291–291, 1949.

[217] J. D. Griffiths. 287. How arbitrary can a constant be? *The Mathematical Gazette,* 57(401):200–201, 1973.

[218] Hunter Watson. A fallacy by parts. *The Mathematical Gazette,* 69(448):122–122, 1985.

[219] Ed Barbeau. Fallacies, flaws, and flimflam. *The College Mathematics Journal,* 21(2):127–128, 1990.

[220] Ed Barbeau. Fallacies, flaws, and flimflam. *The College Mathematics Journal*, 21(3):216–218, 1990.

[221] Ed Barbeau. fallacies, flaws, and flimflam. *The College Mathematics Journal*, 22(2):131–133, 1991.

[222] E. J. Barbeau. *Mathematical Fallacies, flaws, and flimflam*. MAA spectrum. Mathematical Association of America, 2000.

[223] Hugh Thurston. Can we improve the teaching of calculus? *The College Mathematics Journal*, 31(4):262–267, 2000.

[224] Hugh Thurston. A fallacy by parts. *The Mathematical Gazette*, 86(507):443–449, 2000.

[225] Bertrand Russell. Recent work on the principles of mathematics. *The International Monthly*, 4(July–December):83–101, 1901.

[226] C. G. J. Jacobi. De determinantibus functionalibus. *Journal für die reine und angewandte Mathematik*, 1841(22):319–352, 1841.

[227] S. F. Lacroix. *Traité élémentaire de calcul différentiel et de calcul intégral*. Number v. 1 in Traité élémentaire de calcul différentiel et de calcul intégral. Bachelier, Paris, 1828.

[228] Marquis de Condorcet. Mémoire sur les équations aux différences finies. *Histoire de l'Académie royale des sciences (Anée 1770) avec les mémoires de mathématique & de physique*, pages 108–136, 1773.

[229] Marquis de Condorcet. Recherche de calcul intégral. *Histoire de l'Académie royale des sciences (Anée 1772) avec les mémoires de mathématique & de physique*, pages 1–98, 1775.

[230] A. -M. Legendre. Mémoire sur la manière de distinguer les maxima des minima dans le calcul des variations. *Histoire de l'Académie royale des sciences (Anée 1786) avec les mémoires de mathématique & de physique*, pages 7–37, 1788.

Index

Airy, G. B., 153
  Tracts, 219
  (quoted), 37
  Tracts, 153
  symbol for antiderivative, $\int_x$, xxi,
    5, 207
    American users, 224
    Gregory's objection, 223
    Hamilton's critique, 208
algebrification
  of calculus, 159
algebrization
  of calculus, 123
Allendoerfer, C. B. (quoted), xi, 39
Antiderivative, symbol for
  Airy's notation, 5
  in H-Calculus, 8
arbitrary constant, 288
  Heaviside's view, 118

Babbage, Charles, 143
Babbage, Charles (quoted), 74, 151
Bragg, W. H. (quoted), 5
Byerly, W. E.
  (quoted), 12
  books on calculus, 12

calculus
  reform of, xvi
  common conundrums of, xi
  reimport to Britain, 150
calendar
  Gregorian, xviii
  reform of, xviii
chain rule, 16
Condorcet, Marquis de, 298

d^{-1}, 4
∂, 297

adopted by Hardy, 53
in differentiation, 52
in integration, xxii, 80, 299
De Morgan, A., 155
  comments on notation, 208
  critique of Earnshaw's pamphlet,
    192
De Morgan, A. (quoted), xix, 38, 164
de-extinction, xx, 164
derivative, 4, 6
  of a function of a function, 254
determinant
  functional, 62
  Jacobian, 62
differentiability, 251
differential coefficient, 6
differentials
  definition, 14
differentiation
  partial
    symbols for, 54
    several variables, 52
dummy variable, 32

Earnshaw, Samuel
  pamphlet, 167
  theorem, 167
elephants
  species of, xx

Ferrel, William, 225
Fourier, J.-B., 45
Fourier, J.-B. B.
  notation for definite integrals, 45
function
  exponential, 126
  fundamental theorem, 23

Gamow, G. (quoted), 81

323

Hadamard, J. (quoted), 14
Hamilton, W. R. (quoted), 208
Hardy, G. H. (quoted), 6, 52
Heaviside, Oliver (quoted), 118, 123
Hilderbrand, F. B. (quoted), 51
Hobson, E. W. (quoted), 23

integral
    definite, 40
        notation, 45
    indefinite, 8
    particular, 36
    with a variable upper limit, 44
integrand
    in H-Calculus, 9
    in d-Calculus, 4
    in H-Calculus, 9
integration, 7
    by parts, 10, 278, 288, 289
    by substitution, 24
    definite, 23
    indefinite, 23
    inverse of differentiation, 24
    without differentials, 242
$\int_x$ (Airy's symbol for integration)
    De Morgan's critique, 208
    Hamilton's critique, 208
iteration, 84

Jacobi's notation
    Stäckel's comment, 297
    Hardy's comment, 191
Jacobi, C. G. J., 297
    symbol for partial derivative, xxii,
        299
Jacobian (determinant)
    alternative to, 63
    history of, 55
    in thermodynamics, 64
    properties of, 62
    symbol for, 63
Jarrett, Thomas, 205
    notation for derivative, 167

Klein, Felix (quoted), 3, 164

Lacroix, S. F., 297
Lagrange, J.-L

approach to calculus, 162
Landen, John
    algebrification, 159
    residual analysis, 159
Landen, John (quoted), 159
Leathem, J. G. (quoted), xv
Legendre, A. M., 298
Leibniz, 4
Leibniz, G. W. F., xviii
limit
    symbol for, xv

Medawar, P. B. (quoted), 22
Menger, Karl
    calculus without differentials, 241
    symbol for identity function, 240
    variable-free calculus, 238
    view of calculus, 235
Miller, W. H.
    indices, 204
    uses d_x for derivative, 204
Muir, Thomas (quoted), 55

Newton-Leibniz formula, 41, 49
notation
    Bridgman's, for partial derivatives,
        65
    for definite integrals, 45

operator algebra, 119

partial derivatives
    cancellation of differentials, 58
particular integral, 36
Peirce, Benjamin (quoted), 224
Perry, John (quoted), 54
Philip, Alexandeer (quoted), xviii
Phillips, H. B. (quoted), 10
Piskunov, N. S. (quoted), 42
Playfair, John (quoted), 144
Pound, Ezra (quoted), ix, xiv, xvi
principal part, 251, 253
Protter, M. H. (quoted), xvii, 39

Quadling, D. A., 287

Rayleigh, Lord, 118
Russell, Bertrand (quoted), 31, 294

separation of variables, 19
series
    Taylor, 82
series expansion, 82
shift theorems
    for $D + k$, 132
Steller, E. T. (quoted), xii
symbolic method, 123
    protagonists of, 139

Taylor series
    Heaviside's view, 89
theorem
    Earnshaw's, 167
    fundamental, 23, 49
    shift, 132
    Taylor's, 88
Toplis, John (quoted), 144

variable
    active, 53
    change of (in integration), 24
    dummy, 32
    passive, 53
variables
    change of, xxi
    separation of, 19

Weinstock, Robert, 293
Whewell, William, 154
    Doctrine of Limits, 154
Whitehead, A. N. (quoted), 21
Wilson, E. B., 13
    (quoted), 54

www.ingramcontent.com/pod-product-compliance
Lightning Source LLC
Chambersburg PA
CBHW051333200326
41519CB00026B/7410